GEOGRAPHY OF TRANSPORTATION

GEOGRAPHY OF TRANSPORTATION

Commodity Flows and Human Interactions in Meerut City

NARESH KUMAR

Foreword by
AIJAZUDDIN AHMAD

CONCEPT PUBLISHING COMPANY, NEW DELHI 110059

ISBN 81-7022-332-6

First Published 1991

Published and Printed by
Ashok Kumar Mittal
Concept Publishing Company
A/15-16, Commercial Block, Mohan Garden
NEW DELHI-110059 (India)

Lasertypeset by
Microtech Advance Printing Systems (Pvt.) Ltd.
H-13, Bali Nagar
NEW DELHI-110015

Dedicated To My Teacher
Dr. Kanchan Singh

FOREWORD

SPACE ECONOMIES all over the world are organised on the basis of a process of spatial interaction. Human movements and commodity flows are the main instruments through which spatial organisation of economies gets articulated. The process of spatial interaction itself is in response to the forces of economic development. Economies which do not generate surplus remain by and large isolated; while surplus-generating economies forge links through a complex network of exchanges. Within the national space this is the main mechanism through which the domestic market is integrated. The same model applies to international economic order. At the international level economic exchanges between the countries are bound to be unequal since the process of economic development is not uniform. Nor the pressures and pulls within the international economic systems permit a homogeneous process of flow of technologies which lay down the conditions for homogenous development.

Dr. Naresh Kumar examines the question of spatial interaction in the Indian context. His case study of Meerut is symptomatic of the process of economic development in the central doab region of Western U.P. Nestled in the midst of a virile agrarian economy, Meerut has acted as a catalyst of socio-economic change with varying degrees of intensity. Its city region is an ideal illustration of the Dickinson's concept, although Meerut is not a regional city in the strict sense of the term. Nor its linkages are confined to the immediate needs of the agrarian economy of the region it serves.

The city interacts with its surrounding region attracting the surplus produce from the rural and urban centres of collection as it serves as a marketing centre for the agricultural produce. The city also serves as a centre of distribution for commodities which are immediately needed by the rural and urban communities in a surrounding ring of collection centres. The interactive pattern is, therefore, complex and involves inflow and outflow of commodities to and from. In the final analysis, it is the interactive nexus of demand and supply between a hierarchic system of settlements that defines the zone of interaction. Meerut is not a type case by any means and these processes are observed all over India. This work makes a valuable contribution to an understanding of the process of commodity flows in a predominantly agro-based economy. its findings may serve a useful purpose in planning for an integrated development of the market-economy within the country.

Space economy is not autonomous in itself and institutional factors, social structure and variations in accessibility to the surplus generated in different sectors of economy modify the process of spatial integration. Even if the economy is otherwise progressing, it may be leaving behind a section of people who are left-overs in the process of economic development. It is, therefore, desirable to pursue further research with a view to examining the role of institutional and social factors in the process of human interaction and commodity flows. This will be a necessary pre-condition for building a socially cohesive and economically integrated space-economy in a society sharply divided on the basis of caste and religion.

AIJAZUDDIN AHMAD
New Delhi *Centre for the Study of Regional Development*
8 November, 1990 **Jawaharlal Nehru University**

ACKNOWLEDGEMENTS

I HAVE become indebted to a number of persons in the course of this research work. It would have been impossible for me to complete the work without their co-operation and encouragements. As such it is my duty to acknowledge their co-operation earnestly and to express my sense of gratitude to them.

My parents have encouraged me in this work by their blessings, finances and inspirations. This piece of research is, therefore, dedicated to them as a mark of respect. Apart from getting encouragement from my parents, I have received encouragements from my friends and relatives. It is not possible to mention the names, however, I express my sincere thanks to all of them. I have completed this work under the supervision of Dr. Kanchan Singh whose help was available to me at all stages of the work. I express my sincere gratitude to him. I am grateful to my teachers, Dr. N.P. Saxena, Prof. S.C. Sharma, Dr. B.S. Nagi and Prof. Alka Gautam, for their valuable comments, suggestions and encouragements. My thanks are due to Shri Mahesh Chand Sharma and Shri Munshi Lal, who have extended their all possible co-operation and moral encouragement to me.

My most sincere gratitudes are due to the Indian Council of Social Science Research, New Delhi, for its partial financial assistance towards the completion of this work and for its publication. I am extremely grateful to Dr. B.S. Nagi, Incharge, Computer Centre, Council for Social Development, New Delhi, for processing the major exercises. My sincere gratitudes are due to Dr. V.B. Chauhan, Principal, Meerut College, Meerut, for extending me required co-operation from the office and Sri Laxman Dutt, Librarian, for allowing me to use references and other valuable documents in the College Library.

Finally, my thanks are also due to Mr. Ashok Kumar Mittal, Concept Publishing Company, for undertaking the printing and publishing of this book.

NARESH KUMAR

CONTENTS

LIST OF TABLES

LIST OF FIGURES

1

INTRODUCTION

The Problem of Research and the Study Area

The Problem

THE PROBLEM of research revolves round the factor of movement in a given region. The factor of movement involves outflow of surplus produce, inflow of deficit commodities, labour mobility for jobs and purpose-based human interactions. Region is perhaps the first frame that fulfils the aspirations of people and co-ordinates the flow of commodities and human interactions at all levels of settlement hierarchy. This system of movement is not only confined to the regional limits but it moves across, as it depends upon other regions for various demand, supply and services. It is, thus, an inter-connected chain of demand and supply in a hierarchic system of settlements over various regions. Therefore, intra- as well inter-regional connections are essential for any kind of movement. Intensity of interaction (human movement and commodity flows) explains the degree of inter-dependence within a region and among the regions. Hierarchic dependence increases with the increase in the flow of commodities and human movements. To observe the balanced regional growth, it is essential to have advanced linkages (forward as well as backward) and efficient modes of transport. While forward linkages deal with the surplus generation of commodities and creation of additional jobs, backward linkages deal with the adequate supply of raw material, labour, technology and the capital. The flow of commodities and human movements, thus, form an integral part of the nodal relationship.

Interaction is a complex and a dynamic process. Various factors, such as, economic, demographic, social, cultural, political and technological, tend to influence this process. There is a need to analyse the effect of these factors and to estimate the likely prospects of interaction in spatial and temporal contexts. Further, spatial structure of economy needs dispersed chain of production units (agricultural as well as industrial) and service institutions for balanced regional growth. Infrastructure and industrial development remain, by and large, concentrated in urban areas, more so in cities. Since prevailing system of linkages fails to connect all settlements at minimum cost level, regional economy fails to achieve the desired growth levels.

The existing regional structure is, therefore, responsible for growing imbalances between the sectors of economy and areal units. The city draws into itself a large number of rural people and develops into a rapidly growing industrial, commercial and infrastructural centre and, thus, offers opportunities for employment, education, trade, industry, recreation, health, etc. The strategy for urban development should act as catalyst for the improvement of rural environment and balanced development of the city region. There is a need to exploit the potential that lies dormant in rural areas by providing the inputs and the efficient linkages.

The interaction between the urban centres and surrounding rural areas plays significant role in the process of regional development. Interaction between rural and urban areas reflects the inter-dependence among the various units of settlements in a given region. Agriculture is the main sector that provides the livelihood to the majority of the population, trade in agricultural produce is the most important channel accomplishing such an interaction. Another factor responsible for growing interaction is the development of industries. Around these stimuli, develop the necessary infrastructural facilities and amenities which tend to tie hierarchic system of settlements into the chain of dependent units. Inter-regional linkages provide the scope for bridging the deficit gap and for the sale of surplus produce generated in the region.

The Conceptual Framework

The present theme is conceptualised within the framework of space relations whose base is characterised by the regional economy and the top by the interactions. Linkages and mobility are the two axes that connect top with the base. It is, therefore, necessary to define these elements before dealing with their concepts:

Space Relations: Space relations may be defined as the connections between or among the places in a continuous expanse extending in all directions within which all things exist.

Regional Economy: Regional economy is an orderly arrangement of an economic system of the territory or region having almost homogeneous characteristics.

Linkages: Linkages are the system of links organised in a place.

Mobility: Mobility refers to the movement of the people in which passengers move from one place to another.

Interaction: Interaction deals with the act on one another at a given time.

The present theme of research is connected with the concept of space relations. Space relations explain the identity of a place which depends upon the system of regional economy and intensity of interactions. While regional economy is responsible for the level of purchasing power of its people, intensity of interaction is responsible for the degree of dependence among its units (rural as well as urban). The flow of commodities deals with the factors like originating and terminating flows, surplus and deficit commodities, place of origin and destination, institutions to regulate the production, flows and consumptions at various levels. Movement of people, on the other hand, deals with study of purpose, place, frequency and mode of conveyance for inward and outward mobility. Pattern of travel and cost involved in the sale and purchase of the goods decide about the choice of the interaction for a place.

The Study Area

There are two inclusive parts of the study problem and the study area, viz. (i) significance of interaction and (ii) spatial influences on the interaction. The first part of the study relates to the Meerut City. It covers the chapters on spatial and temporal flow of commodities, case studies, spatial and temporal movement of population and correlates of interaction. The second part of the study area relates to the district

Meerut. It covers a section on the functional gaps and income inequalities in Chapter 2, determinants of interaction in Chapter 6 and problems and prospects of inter-action in Chapter 7. Thus, first part of the study deals the locational importance of Meerut City in terms of interaction whereas second part deals with the city within a district frame where spatial influences are analysed to understand the relative position of the constituent units (18 blocks). Details about Meerut city and Meerut district are given below (Fig. 1.1):

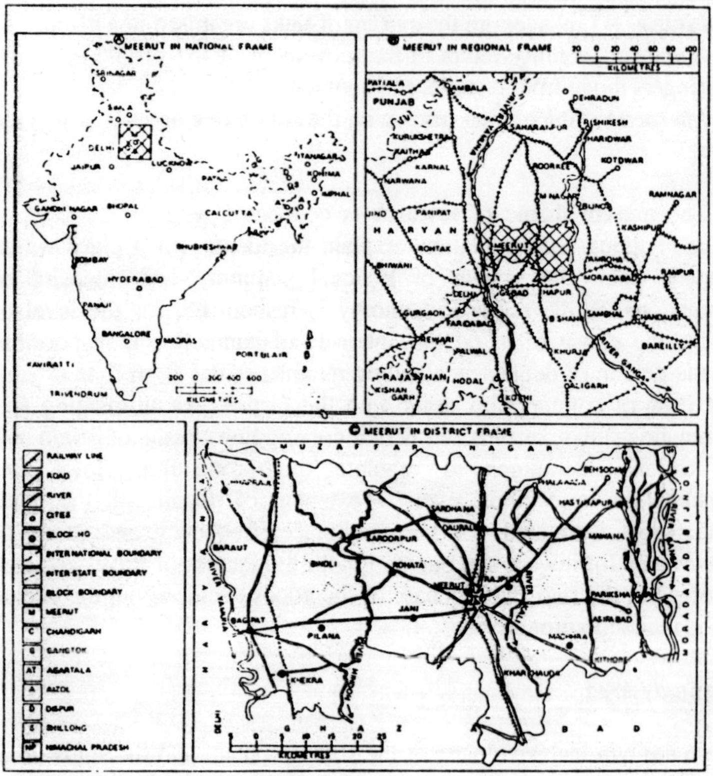

FIG. 1.1: Location map of Meerut.

Meerut City

Meerut city lies at an inter-section of 28° 58′ N and 77° 42′ E. The

1981 Census registered a population of 5,36,615 persons spread over an area of 80.82 sq. km. There were as many as 1,01,222 houses and 1,01,584 households during 1981. Meerut city registered an annual addition of about 17,000 persons, about 5,000 households and an extension of over 2.5 km of area in the city limits. An annual growth rate of 4.59 per cent and a density of 6,640 persons per sq. km was recorded during the period under study. In terms of interaction, there was an annual turnover of 2.93 million tonnes of commodities and 19.56 million passengers during 1984. Of the total turnover of the commodities, nearly 20 per cent was recorded as outflow and remaining 80 per cent as inflow. The annual earnings, through movement of passengers, reached the level of Rs. 106.08 million by 1984. *Mandi* arrivals in Meerut weighed about 1.54 lakh tonnes, milk arrival about 20.82 thousand tonnes and textile arrivals 5.04 lakh metres. Of the total outflow, nearly 7 (7.45) per cent moved through railways remaining 93 (92.55) per cent through highways. Similarly of the total inflow, 28(28.11) per cent arrived through railways and remaining 72(71.89) per cent through highways. While outflow of cotton textile was about 165 million metres worth Rs. 806.36 million, outflow of sports goods was 1,854.63 tonnes worth Rs. 73.48 million and of cutlery goods 87.03 tonnes worth Rs. 6.1 million. Human interaction covered up to Bombay and Nagpur in the South, Mugalsarai and Saharsa in the East, Jammu Tawi and Atari in the North-West and Ajmer in the South-West. Unlike long distance interactions through railways, highways recorded a limited range of interactions, i.e., moderate distances only up to 400 km. Human interactions, through highways, included places, such as, Chandigarh, Dehradun and Rishikesh in the north, Haldwani and Tanakpur in the east, Lucknow, Lakhimpur and Palia in the South-East and Agra in the South. While Meerut City recorded zonal (northern as well as western zone) space relations through human movements, it recorded national space relations through commodity flows.

District Meerut

District Meerut lies between 28° 52′ N to 29° 18′ N and 77° 07′ E to 78° 12′ E. There are 4 tehsils, 18 development blocks and 913 inhabited villages in this district. According to 1981 Census, Meerut district had a population of 27,67,246 persons. About 69 per cent of the population, in the district, lives in rural areas and rest of it (about 31%)

resides in urban areas. Population growth over a decade (1971-81) was 25.33 per cent. However, growth of rural population was only 11.42 per cent whereas that of urban population was 72.87 per cent. Proportion of literate persons was 34.68 per cent during 1981. Male literacy accounted for 46.73 per cent whereas that of female literacy for about 20.30 per cent. Proportion of literacy among rural as well as urban areas was 30.01 and 44.97 per cent respectively. The proportion of main workers to total population was 27.68 per cent, of marginal workers 0.61 per cent and of non-workers 71.71 per cent. The district covers an area of 3,92,191 hectares of which 3,09,583 hectares (79%) area is under cultivation. Of the total area under cultivation, over 90 per cent is covered by the different sources of irrigation. The average size of the land holding in the district is 1.4 hectares. About 77 per cent of the total holdings is less than two hectares in size. The district livestock population was 13.15 lakhs according to 1977 livestock census. Of the total livestock, number of cows and buffaloes was 1.42 lakhs.

The big factories in the district accounted for a sum of Rs. 110 crores and for the employment of 11.6 thousand persons. There are 19 such factories in the district. In addition to that three more big factories are under construction. There are about 3,143 registered small scale industrial units, besides 5,572 unregistered units, in the district. Of these, only 238 units are registered under Factories Act.

In terms of services, about 86.52 per cent villages are provided with education, 20.22 per cent with medical, 28.70 per cent with post and telecommunication, 10.98 per cent with the market, 38.15 per cent with communication, 63.59 per cent with approach roads, 90.22 per cent with power supply and cent per cent with the supply of drinking water facilities. The per capita consumption of electricity in the district amounts to 150 kWh as against 87 kWh for the State and 130 kWh for the nation as a whole.

These facts explain that Meerut City has a developed commercial and industrial base for the interaction. Besides growth generative base, it has a fairly large number of services and institutions (like education, medical and administrative) which add to the scale of interaction with the city. Apart from developed urban economy, Meerut has also developed rural economy which acts as the source of surplus flow to the city. Thus, advanced economic base of the region has established forward linkages with other locations in India.

Review of the Literature

The study of the available literature on the present theme may be grouped under four heads for the convenience of interpretation. Studies are related with:

(i) Commodity flows,
(ii) Human movements,
(iii) Information flows and
(iv) Rural-urban relationship.

Following is the sequence and scope of studies conducted so far under each group. Chronological profile of contributions is followed by the sequential growth of the literature. References are cited along with the individual reviews in the text.

Commodity Flows

The study of commodity flows formed an important aspect of geographical researches right from '50s. Contributions of Alexander (1958). Berry (1966), Linneman (1966), Reed (1967), Ilori (1968), Chatterjee (1969), Pankaj (1968), Hodder (1969), Hay and Smith (1970), Coppock (1971), Chisholm (1971, 1973), Adalemo (1975), Gezann (1975), Wanmali (1975), Pal (1977), Raza (1980), Kara (1982), Rao (1986) and Baer and Birch (1987) are noteworthy in this regard. Alexander's[1] study on United States of America makes an evaluation of comparative freight rates between uniform and nodal regions. Results reveal that freight rates were minimum in case of nodal regions, due to high connectivity and available alternatives. Linneman[2] studied exports and imports of Holland. With the help of input-output model, structure of international trade was analysed. Hodder and Ukwu[3] analysed the trade among the Yoruba and Ibo tribes of West Africa. Study deals with the flow of commodities in a tribal economy. Nigeria's structure of inter-regional trade and transfer of money was analysed by Hay and Smith.[4] Study deals with the surplus and deficit flows within a national frame-work. Adalemo[5] studied the marketing behaviour of Nigerian centres. The study covers the aspects of travel behaviour of traders, circular causation of marketing rings and patterns of market shifts. Observations reveal that periodic markets play a vital role in meeting the local demands (up to 65 %) in an under-developed set up like that of Nigeria. Agricultural crops, livestock and enterprise combinations were analysed for England and Wales by Coppock.[6] This study brings out how three factors are inter-dependent causing inter-regional flows in the U.K. British economy was further analysed by Chisholm.[7] This study covered the aspects of freight transport costs, industrial location and regional development. With the help of freight flows, regional patterns were identified and a model was developed to explain the present frame of British economy. Gezann[8] analysed itinerant traders and sale of commodities in Lushoto periodic market of Tanzania. This study deals with movement of traders from one periodic market to the others for selling their commodities. Baer and Birch[9] attempted a survey on the international economic relations of Paraguay. Study analyses the international economic relations of the Paraguay and its export-import relations with the other countries of the world. It is concluded that economic development does influence cultural change though both the factors are not co-terminus. To achieve higher rates of economic development, it is essential to develop a growth generative economy by raising the purchasing power of its people.

The study of commodity flows received scant attention by scholars in India. The first attempt to analyse the spatial structure of Indian economy was made by Berry[10] in 1966. This study used railway information about the movement of various commodities over different parts of the country. Commodities were grouped and were plotted on the maps to analyse the extent of surplus generations and deficit demands. Results reveal that India's international trade is port oriented. A year later, i.e., in 1967 Reed[11] worked out the nature of commodity flows from Bengal-Bihar industrial belt. The observations reveal that industrial development accelerates flow of commodities and thus helps in the process of regional development. Chatterjee,[12] with the help of empirical observations, discussed the nature of commodity flows from the fringe areas of Calcutta city. Commodities selected for the analysis were vegetables, fruits, fish, rice, pulses, milk and milk-made items. Pankaj[13] undertook the survey of commodity flows to and from the Cochin port. With the help of chart (inflow and outflow) and flow gradient, it was possible to de-limit the hinterland limits of Cochin port. Results reveal that port of Cochin received most of its supplies from the States of Kerala and Karnataka. Wanmali and Ghosh[14] worked out the patterns and prospects of goods and services in south Bihar. Exercise is based on the labour mobility for various services and inter-centre flow of commodities in the region. It was possible to workout the likely trend of interaction. Study concluded that social and political factors worked as constraints in the absence of adequate facilities and services. Pal[15] estimated the prospects for different commodities based on the factor of regional demand. Study refers to the period of 1970-71 and 1975-76. Rate of change and deficit demand of the regions were the main criterion used for the projection in this exercise. It was estimated that regional demand of commodities would increase twice to its present level within a period of 10 years. Raza[16] carried out a detailed investigation of the structure of commodity flows in India. Movement of commodities through highways and railways were analysed separately. Finally, regression equations have been worked out to explain the effect of distance, weight, freight, mode and connectivity. It was concluded that structure of commodity flows still follows the colonial system of linkages and it fails to connect the resource points in the interior areas at the minimum cost level. Its basic functional character, however, remains unchanged. It continues to be the focal point for commodity, income, capital and labour force flows. Traffic characteristics of major Indian ports were studied by Kara *et al.*[17] Researchers have worked out specialisation index for each group of commodities moving in an out from these ports. Backward and forward linkages were analysed and inter-port variations were discussed. About 50 per cent of the total flow of commodities related to the minerals and petroleum products. Besides these flows, there is an increasing tendency of balance of trade among the various ports. Over 50 per cent of tonnage related to the foreign trade. Coastal trade is not, therefore, so developed. There is a need to increase the coastal linkages to develop these areas. Rao[18] studied the nature of regional economy of Andhra Pradesh. Study focused upon the structural changes introduced in the regional economy. Having analysed the performance of various sectors, inter-sectoral linkages were analysed. Results reveal that regional economy of Andhra Pradesh is reeling under growing constraints. However, coastal Andhra has experienced relatively higher rate of substitution compared to Telengana and Rayalaseema. Inter-sectoral shifts in a regional economy were possible only through developed linkages.

Human Movements

Studies on this aspect of interaction have been conducted by Berry,

Barnam and Tennant (1962), Kansky (1963), Olsson (1965), Murdie (1965), Chapman (1970), Goddard (1970), Singh (1971), Wilson (1971), Ramu and Wiebe (1973), Harvey, Hocking and Brown (1974), Mitchell (1978), Kulkarni (1981, 1984), Bose and Bhattacharya (1984), Kumar (1985) and Huff and Hanson (1986).

Berry *et al.*[19] analysed the consumer travel pattern with reference to the retail trade locations. Locational pattern of retail trade reflects the behaviour of consumers. The order of a commodity and its market competition determines the flow of consumers and their preferences for various goods. Murdie[20] analysed consumer travel behaviour in relation to the cultural setting. It was concluded that in a plural society, preference for a place is determined by its cultural base. Cultural base is decided by the intensity and cultural combinations at a given place. Patterns of such movements are common in areas having relatively low socio-economic advancement. However, in a modern society, place preference and intensity of human movements are explained by the level of cultural awareness and economic advancement. Ramu and Wiebe's[21] study on mobility behaviour is a unique as it limits the operation to the two components only, i.e. (i) Occupational and (ii) Educational mobility. The main thrust of the study was to analyse the role of caste as an institution in India. The effect of caste on human mobility was analysed under the groups of Scheduled Caste, backward caste and upper caste.

Transport network is a pivotal element accomplishing human interaction. Kansky[22] analysed relationship between network geometry and regional characteristics. Network geometry explains regional characteristics in terms of human movement, its preferences, frequencies and modes. Purpose of movement, however, varies significantly in case of place preferences. Olsson[23] reviewed the concept of distance and human interaction in global context. Results confirm the finding that interaction is explained by the factor of distance. However, with the new means of transport, technology and alternate modes of transport, effect of distance is minimised considerably. Harvey *et al.*[24] analysed relationship between the consumer travel patterns and the behaviour of the salesman. Based on the growing pressure of population and chromatic travel pattern, an exercise has been attempted to project the population and plan the geographic space for future use. Kumar[25] analysed the economic inter-dependence of Meerut city with surrounding rural places and urban centres in other parts of the country. Besides analysing the economic inter-dependence, movements of people (to and from Meerut) through highways and railways have also been evaluated. Results reveal that the complementary effect of Delhi metropolis has raised the extent of interaction on Meerut-Delhi axis. As a push effect of Delhi, Meerut is experiencing fast urban growth and increase in the quantity of commodity flows. Surplus flow of commodities is due to advanced agricultural as well as industrial base of the regional economy. Bose and Bhattacharya[26] analysed the growth trends of Mysore city and measured the impact of urban growth on the passenger transport. Observations reveal that passenger transport falls short of the demand of the fast growing urban population of Mysore city. It is, therefore, suggested that planned attempts must be made to increase the level of passenger transport to meet the demand of the city.

Goddard,[27] with the application of factor analysis, studied functional regions within a city centre. Taxi flows were considered as the most suitable indicators in explaining the functional gravity of a place within a city. Chapman[28] worked on tribal interactions which relate to the Duidui and Pichahila tribes of British Soloman Islands. It has been observed that human movements were very limited in a tribal society. However, frequency increases on certain occasions and on specific seasons. Wilson's[29] study deals with the family interactions. Based on the temporal observations, a model has been developed to explain the family interactions. Singh[30] discussed the rail traffic flow in Rajasthan. It has been observed that passenger traffic is increasing along the eastern part of the Rajasthan whereas it remains considerably low in the western part of the State. The central part of the State presents moderate traffic movement. Uneven rail traffic is partly due to varying connectivity and partly due to varying productivity of the State economy. Mitchel[31] discussed the issue of mobility in relation to wage-labour variations. Prevailing structure of wages and pattern of employment are the major determinants of labour mobility at a given place and time. Diurnal as well as seasonal mobility of labour depends upon the capacity of job generation and surplus production. Kulkarni[32] analysed the functional structure of Nasik city. Functional gravity does not remain uniform over all parts of the city. Consequently, human movement varies from place to place. Movement for a variety of purposes ranging from wage-labour to recreational and shopping are caused due to the diversity of functions in a city. Kulkarni[33] made another study of Ahmedabad city. In this study, he analysed the factors behind rapid crowding in Ahmedabad city. Unplanned urban growth has led to the congestion and crowding in certain localities of city. Finally, an integrated model for urban development has been suggested to control the growing congestion in the city. Huff and Hanson[34] calculated the possibility of repetition and variability in urban travel. Based on the city survey of Ohio it was concluded that urban travel is for a variety of purposes and has high chances for the repetition. It is further observed that variations are large both on the travels over different places in the city and on the travels over the period of time.

Information Flows

Flow of information forms the major source base for spatial integration at all levels of hierarchy (national, regional and local). It helps in bridging the existing information gap and raising the level of awareness. As such information flows are the important aspect of interaction studies. Tornqvist[35] worked out the degree of dependence and extent of relationship between information flows and economic activity. Such flows include agricultural, industrial, educational, legal and recreational informations. Economic activities such as primary productions (agriculture, pastoralism, hunting, fishing and lumbering), processing (household industries, manufacturing, transport) trade and services are affected by the flow of informations. Results reveal that information flows have played vital role in accelerating the economic activity of a place. Spread of information depends upon the efficient media network and means of communications. Pred and Tornqvist (1973)[36] studied nature of information flows and the systems of the cities. System of hierarchic dependence increases with the diffusion of information among cities. Results reveal that flow of information does not trickle down in a hierarchic manner but follows the order of linkages. Physical barriers and social taboos affect the information flow adversely. Gillespie and Goddard[37] analysed the role of telecommunications in regional economic development. Results reveal that the production technology determines the level of economic development in a region whereas technology of telecommunication determines the economic integration among the regions.

Rural-Urban Relationship

Study of rural-urban relationship forms an important basis for geographic researches in India and abroad. Contributions of Morril

(1965), Olsson (1965), Clark (1967 & 1968), Wanmali (1972), Holmes, Williams and Brown (1972), Alam and Khan (1972), Wanmali *et al.* (1975), Gopi (1978, 1980), Deshpande, Arunachalam and Bhat (1979), Rao and Tewari (1980), Mortimore (1981) and Banerjee (1987) are important in this field of research. Growing concentration of economic activities in urban areas have accelerated rural migration towards urban areas for jobs and other economic gains. Consequently, there has been a rapid growth in urban population. Morril[38] analysed migration and consequent growth of urban areas. Study relates to the three sets of migration, viz., short, medium and long distance migrations. Finally distance slabs were correlated with the type of migrants, viz., rural-urban and urban to urban. Olsson[39] reviewed the question of distance in human interaction. Clark[40] worked out the concept of range in the interaction. The consumer travel patterns formed the basis for this concept. Results reveal that short distance movements are extremely frequent and frequency of movement decreases with the distance. Wanmali[41] worked out the provision of social facilities in a region following the technique of scalogram. The distribution pattern of facilities provided the base for the identification of functional as well as areal gaps. Finally, a scheme of planning has been suggested to remove the gaps in the distribution. Chapman and Wanmali[42] analysed the rural-urban relationships in India. Potential regional demands and population estimates have been worked out with the help of projections. Alam and Khan[43] surveyed the Hyderabad metropolis to assess the rural-urban relationship. Having analysed the major city problems, an integrated plan for balanced regional development has been suggested. Holmes *et al.*[44] estimated the pattern of travel under maximum restrictions. It is concluded that a location with a distinct facility set up generates interaction despite restrictions posed over it. A model, based on the lines of creche – the day-care facility – has been suggested for balanced urban growth. Distribution pattern of consumer goods in rural India has been studied by Wanmali and Ghosh.[45] The authors have assessed the feasibility of the central place theory for the distribution of consumer goods in an Indian frame of development. Facts reveal that social and political factors tend to influence the pattern of goods and services in rural India. Location-allocation models were tested to develop the least cost matrix for rural-urban interaction. Gopi[46] attempted an exercise on the delimitation of urban fringe and analysed the broad characteristics of these areas. Rural territory surrounding urban areas gradually transforms its character due to urban encroachment. A model has been developed to explain the characteristics of urban fringe. Deshpande, Arunachalam and Bhat[47] carried out an interaction survey of south Kolaba. Study explains the nature of rural-urban relationships in a hierarchic system of settlements. Due to the pull effect of Bombay metropolis, interaction along Bombay-Kolaba axis has increased considerably. Interaction has been analysed within the frames of purpose,. place, frequency and modes of conveyance. Finally, the system of settlements has been correlated with the extent of interaction to explain the dynamics of regional economy that functions variously over five levels of settlement hierarchy. Prakasa Rao and Tewari[48] analysed the structure of an Indian metropolis. A detailed urban survey of Bangalore was carried out for this purpose. Economic structure of the metropolis is discussed under the groups of commercial, industrial, housing and service sectors. Further, nodality of the metropolis has been worked out on the basis of rural-urban inter-dependence and regional dominance (primacy) in the space economy. Finally, number of projections have been worked out to suggest a plan for the metropolis. It has been observed that massive migration to the metropolis poses most serious challenge to the regional development. As such a check to this migration is urgently needed. It has been suggested that the creation of jobs and facilities at the small and medium sized towns will have positive impact upon the re-distribution of population. Mortimore[49] developed an alternate opportunity model for population mobility in Nigeria. The model explores the possibilities of the migration towards rural areas instead of urban areas. Planned efforts are suggested to gear up the production technology in rural areas to check the migration at source. Diversification in the rural economy and its developed links with the market are likely to create balanced conditions for rural-urban relationships. Benerjee[50] analysed rural-urban migration in case of Delhi metropolis. Rural migration to Delhi is from short as well as long distance zones. Long distance rural migration is from the areas of western Bihar, eastern and southern Uttar Pradesh, north-western Madhya Pradesh, northern Rajasthan and western Haryana. Rural migrants are employed in informal sector, housing, hoteling and construction works of the metropolis. Besides these migrants, a sizable section of the labour supply is from neighbouring rural areas, i.e., from the range of 80 km. Diurnal mobility into urban labour market is for a variety of industrial, commercial and service jobs, diurnal labour mobility, thus, fulfils the labour demand of Delhi.

The foregoing literature survey reveals that studies on commodity flows were taken up with a view to explain the space relations at three levels, i.e., (i) international trade, (ii) national economy and nature of commodity flows and (iii) regional aspect of flow taking formal or functional regions into consideration. Alike commodity flows, human movements have also been analysed from various perspectives, i.e., cityward migration, consumer travel pattern, work place and functional mobility. Most of these studies deal with the regional dimensions of human movement. Similarly, rural-urban relationships have been studied with the factors, like, distance, purpose, place, frequency and mode. Provision of service facilities and trade forms the basis of centrality for such a relationship.

It is, therefore, clear that there have been limited studies dealing with the space relations of a city in terms of feed-back to rural areas and flow of regional and local surplus to inter-regional and international centres. Similarly there has been no systematic study that deals with intra-regional as well as inter-regional human movements with reference to a nodal centre. Studies on rural urban relationships have not adequately explained the integral functional link of surplus and deficit commodities. As such a detailed study is needed to unfold the backward and forward linkages of the city in terms of commodity flows and human movements.

A case study of Meerut presented here covers most of these aspects. Being a developed region both in terms of agriculture as well as industry, it has huge quantity of surplus from both the sectors of production. Most of the human movement is directed towards Delhi – the nearest metropolis. Meerut serves the Delhi metropolis with the supply of consumer goods, surplus goods and labour. Interaction in this part of the country is, therefore, growth generative. As such it is essential to analyse the rural to urban and urban to urban relationship to understand the dynamics of regional growth. In the process of regional development city acts as a focal point of interaction influencing its surrounding country-side and other cities in the country.

Objectives of the Study

The objectives of the study are as follows:

(i) to study the role of Meerut city as a nodal centre in regional and national perspectives by analysing the spatial and temporal trends of commodity flows and human movements,

(ii) to analyse the various correlates of interaction and to determine the role of linkages, technology and institutions in the process of interaction,

(iii) to assess the problems and prospects of interaction.

Methodology

Sources of Data

The study is based upon both secondary and primary sources of data. The secondary data were collected from the office records of railways, municipality, *mandi parishad* and highway depots. Informations about industrial production, income, employment, investment, etc., were collected from the office of the General Manager, District Industries Centres, Meerut. Directory of Industries, District Meerut, 1984, proved to be the useful document about such informations. Data regarding population, agriculture and infrastructure were taken from the Statistical Abstract – District Meerut, 1984. Informations about poverty survey, poverty eradication programme and financial supports to the rural areas were collected from the District Rural Development office, Meerut. District Annual Plan, 1986-87 formed the source for data regarding the financial allocation of various sectors of economy and of various areal units (18 blocks). Informations about newly constructed roads, approach roads and other rural roads were collected from the office records of P.W.D., Meerut. Details about the modes of transport, their numerical strength in different blocks and routes and the nature and types of permits issued were collected from the office of the Regional Transport Officer, Meerut. Data regarding production, flow relations and prices of the cotton textiles were collected from the offices of the District Industries Centre, Meerut, Assistant Director, Industries (Handlooms), Meerut Division, Hathkargha Bhawan, Meerut and Weaver's Service Centre, Meerut. Statistics of arrival, supply and production of milk and milk-made items were collected from the office of the Feeder Balancing Dairy, Partapur, Meerut. Apart from the collection of data from various offices, necessary informations were also obtained from offices of the various manufacturing associations such as sports goods, handlooms and cutlery.

It is, therefore, evident that large part of the book is based on the secondary data. However, certain informations which were not available with these offices, were collected through field survey. Field survey was conducted during the months of May, August and November 1986, to collect primary data. The selection of sample months is based on the effect of seasons and crop harvests upon the movement of people and the commodities. While month of August represents rainy season, months of November and May represent winter and summer seasons, respectively. Mobility survey was conducted at seven places viz., taxi stand, bus stands and railway stations. Two hundred and twenty-five sample passengers were interviewed for their place of origin and destination, purpose of interaction, frequency of visit and modes of conveyance used. Besides survey on population mobility, industrial surveys were also conducted in cases of three industries, viz. cotton textiles, sports goods and cutlery. Industrial survey covered the enquiry about raw material supply, amount of production, distribution, industrial problems and prospects. In all 85 sample schedules were filled up. Of these, 40 samples were from cotton textiles, 30 from sports goods and 15 from cutlery. For cotton textiles, field survey was conducted in Khandak and Golakuan localities, for sports goods in Surajkund and sports goods complex in Partapur localities and for cutlery in Sabungran locality. Thus, apart from obtaining secondary data, specific interviews of 310 samples helped in the completion and compilation of the required data.

Variables

Variables bridge the gap between the vague idea and definite conclusions. Evaluation and interpretation of variables is as such very essential for generalisations and conclusions. Interpretations are made and inferences are derived through the distribution of variables over time and space. Quantity of commodity arrival and outflow has been used as variables to explain the spatial and temporal flow of commodities. While flow differentials have been used as a variable explaining surplus and deficit commodities, industrial concentrations and income inequalities have been used as variables explaining the functional gaps and income imbalances. Variables, such as, raw material supply, quantity of production and outflow have been used to discuss the industrial case studies (cotton textiles, sports goods, milk and cutlery). Intensity of place, purpose, frequency and modes of conveyance have been used as variables to explain the spatial and temporal movement of the population. Correlates of commodity flows and human movements have been explained through variables like market prices, demand, production and distance. Technological, transport and communication and institutional variables have been used to explain the determinants. Similarly, geographical, social and economic variables were used to discuss the problems whereas agricultural, non-agricultural development, I.R.D.P. and Poverty Eradication Programmes and budget allocations were used to discuss the prospects of interaction.

Hypotheses

The proposed theme of research seeks to evaluate the following hypotheses:

(a) Status of the regional economy is reflected in the surpluses generated and deficits demanded in a region,
(b) Intensity of interaction is proportional to the nodality of a centre,
(c) Intensity of linkages explains the extent of space relations, and
(d) The amount of interaction is inversely proportional to the distance.

Methods Used

Statistical and cartographic methods have been used in the research. Cartographic methods of proportional circles, bar, linegraph, traffic flow diagrams, isopleth and choropleths have been used to explain the spatial dimensions of interaction. Statistical methods of ratios, proportions, concentration, flow differential, functional gaps, standard deviations, index numbers, normalised values, composite scores, correlations and regressions have been used to measure the commodity flows and human interactions and to estimate the prospects of overall interaction. Multivariate and complex statistical exercises were processed at the computer centre, Council for Social Development, New Delhi.

Scope and Plan of the Study

Present theme of research has wide scope of enquiry. Beginning from the aspects of commodity flows and human movements over time and space, it leads to the identification of problems and prospects. The entire exercise passes through the complex but inter-connected variables. Complexity of the variables remains all through the text, i.e., from the comparison of the data to their responses and prospects. Normalisation of the variables and conversion of actual values into index numbers have made the data comparable and addable. With the help of case studies, sample surveys and by working out responses, determinants and problems, nature of interaction and its prospects have been explained.

The study is organised into eight chapters. Chapter 1 deals with the statement of the research problem, study area, review of the literature, the conceptual framework, methodology and scope and plan of the study. Next, to the Introduction, is the chapter 2 on spatial and temporal flow of commodities. It deals with the arrival and outflow of commodities, nature of surplus and deficit commodities and functional gaps and income inequalities. As a supplement to this chapter, case studies on cotton textiles, sports goods, cutlery and milk have been analysed in Chapter 3. Movement of population in spatial and temporal frames have been studied in Chapter 4. Discussion in this chapter revolves round the factor of place, purpose, frequency and modes of the movement. Correlates of the commodity flows and human movements have been analysed in Chapter 5. Price, demand, production, distance, purpose, frequency and mode of conveyance have been used to explain the extent of interaction. Having worked out the correlates (commodity flows and human movement), determinants of interaction have been analysed in Chapter 6. Determinants have been grouped into technology, transport, communication and institution. The extent to which a determinants explains interaction has been analysed separately as well as collectively. Chapter 7 is devoted to the assessment of problems and the prospects of interaction. While problems have been analysed under the dimensions of geographic, social and economic; prospects have been analysed under the dimensions of agricultural and non-agricultural developments and I.R.D.P., Poverty Eradication Programmes and budget allocations. Finally, the results of the study and conclusions have been discussed in Chapter 8. The appendices and questionnaire schedules that formed the basis for normalised tables and discussions in the text are also given in this book. The text of book is supported by a comprehensive bibliography covering the latest articles and books.

NOTES

1. J.W. Alexander, (1958) "Freight Rates : Selected Aspects of Uniform and Nodal Regions". *Economic Geography*, 34, pp. 1-18.
2. H. Linneman, (1966) *An Econometric Study of International Trade Flows*, North Holland.
3. B.W. Hodder and U.I. Ukwu, (1969) *Markets in West Africa: Studies of Markets and Trade among the Yoruba and Ibo*, Ibadan University Press.
4. A.M. Hay & R.H.T. Smith, (1970) *Inter-regional Trade and Money Flows in Nigeria*, Oxford University Press.
5. I.A. Adalemo, (1975) "Traders Travel Patterns : Marketing Rings and Pattern of Market Shift". *Nigerian Geographical Journal*, 18.
6. J.T. Coppock, (1971) "Crops, Livestock and Enterprise Combination in England and Wales", *Economic Geography*, 40, pp. 65-81.
7. M. Chisholm, (1971) "Freight Transport Costs, Industrial Location and Regional Development", in M. Chisholm and G. Manners (eds.), *Progress in Geography.*
8. G.A. Gezann, (1975) "Periodic Markets and Itinerant Trader's Commodities in Lushoto, Tanzania" in Robert H.T. Smith (ed.) *Internal Exchange Systems : Periodic Markets and Hawkers in Africa, Asia and Latin America*. Melbourne: Sorrett.
9. Werner Baer & Melisa Birch, (1987) "The International Economic Relations of a Small Country : The Case of Paraguay", *Economic Development and Cultural Exchange*, 35, (3), The University of Chicago Press, April, pp. 601-27.
10. B.J.L. Berry, (1966) "Essays on Commodity Flows and the Spatial Structure of the Indian Economy", *Research Paper No. 109*, University of Chicago, Department of Geography.
11. W.E. Reed, (1967), "Areal Interaction in India : Commodity Flows of the Bengal-Bihar Industrial Area", University of Chicago, Department of Geography, *Research Paper No. 110*.
12. A.B. Chatterjee, (1969), "Nature of Commodity Flow of a Market in the Fringe Area of Calcutta", *Geographical Review of India*, 31, (3).
13. T. Pankaj, (1968), "A Study of the Hinterland Limits and Traffic Flow Patterns of the Port of Cochin", *Arthavijnana*, 10.
14. S. Wanmali and Avijit Ghosh, (1975) "Distribution of Goods and Services in South Bihar : Patterns and Prospects", in Hauser, Waltar and Manor, James (eds.). *Two Faces of India, Social and Political Change in Bihar and Karnataka*, Oxford University Press, New Delhi.
15. M.N. Pal, (1977) "Regional Demand Projects for Different Commodities in India 1970-71 and 1975-76", *Indian Economic Journal*, 14 (4).
16. M. Raza, (1980) "Structure of Commodity Flows in India". *Project Report* Submitted to the Planning Commission of India.
17. P.K. Kara, S.N. Tripathy and D.K. Singh, (1982) "A Comparative Analysis of Major Ports of India and their Traffic Characteristics", *N.G.J.I.* 28, Part 1&2, Varanasi, pp. 60-66.
18. K.P.C. Rao, (1986) "Structural Changes, Sectoral Performance and Inter Sectoral Linkages in Andhra Pradesh Economy", *Margin*, 18 (2) Jan. pp. 62-69.
19. B.J.L. Berry, H.G. Barnum and R.J. Temant, (1962) "Retail Location and Consumer Behaviour", *Papers and Proceedings of the Regional Science Association*, pp. 65-106.

20. R.A. Murdie, (1965) "Cultural Differences in Consumer Travel", *Economic Geography*, 41, pp. 211-33.
21. G.N. Ramu and P. Wiebe, (1973) "Occupational and Educational Mobility in Relation to Caste in Urban India", *Sociology and Social Research.* 58 (1).
22. K.J. Kansky, (1963) "Structure of Transportation Networks : Relationships between Network Geometry and Regional Characteristics," *Research Paper No. 84*, University of Chicago, Department of Geography.
23. G. Olsson, (1965) *Distance and Human Interaction : A Review and Bibliography*, Regional Science Research Institute.
24. M.E. Harvey, R.T. Hocking & J.R. Brown, (1974) "The Chromatic Travelling Salesman Problem and its Application to Planning and Structuring Geographic Space". *Geographical Analysis*, 6, pp. 33-52.
25. N. Kumar, (1985) "Economic Interdependence and Human Interaction in Meerut City", *Proceedings of the Sixth N.A.G.I. Seminar*, I.I.T., Kharagpur, 9-11 May, pp. 137-38.
26. P.K. Bose and R.N. Bhattacharya, (1984) "Growth of Mysore City and its Passenger Transport", *The Geographical Review of India*, 46, (4), Dec. pp. 34-45.
27. J.B. Goddard, (1970) "Functional Regions with a City Centre : A Study by Factor Analysis of Taxi Flows in Central London", *Transactions of the Institute of British Geographers*, 49, pp. 160-80.
28. M. Chapman, (1970) *Population Movement in Tribal Society : The Case of Duidui and Pichahila, British Salomon Islands.* Ph.D. dissertation (Pub.), University of Washington.
29. A.G. Wilson, (1971) "A Family of Spatial Interaction Models and Associated Developments", *Environment and Planning*, 3, pp. 1-32.
30. I.D. Singh, (1971) "Pattern of Rail Traffic Flow in Rajasthan", *National Geographical Journal of India*, 17, (2 & 3), Varanasi, pp. 90-98.
31. J.C. Mitchell, (1978) "Wage Labour Mobility as Circulation" : Paper Presented at the International Seminar on *The Cross-Cultural Study of Circulation*, East-west Population Institute, Hawaii.
32. K.M. Kulkarni, (1981) *Urban Structure and Interaction : A Study of Nasik City Region*, Inter-India Publications, New Delhi.
33. K.M. Kulkarni, (1984) *Geography of Crowding and Human Response : A Study of Ahmedabad*, Concept Publishing Company, New Delhi.
34. James O. Huff and Susan Hanson, (1986) "Repetition and Variability in Urban Travel", *Geographical Analysis*, 18, (2), Ohio State University Press, April, pp. 97-114.
35. G. Tornqvist, (1968) "Flows of Information and the Location of Economic Activities", *Geografiska Annaler*, Ser. B, pp. 99-107.
36. A.R. Pred and G.E. Tornqvist, (1973) "Systems of Cities and Information Flows", *Lund Studies in Geography*, Ser. B, 38.
37. A.E. Gillespie and J.B. Goddard, (1986) "Advanced Telecommunication and Regional Economic Development", *The Geographical Journal*, 152 (3), Nov. pp. 383-97.
38. R.L. Morril, (1965) "Migration and Growth of Urban Settlements". *Lund Studies in Geography*, Series B, Human Geography, (26), pp. 130-70.
39. G. Olsson, (1965) *Distance and Human Interaction : A Review and Bibliography*, Regional Science Research Institute.
40. W.A.V. Clark (1968) "Consumer Travel Patterns and the Concept of Range".

Annals, Association of American Geographers, 58, pp. 386-96.
41. S. Wanmali, (1971) *Regional Planning for Social Facilities*, National Institute of Community (now known as rural) Development, Hyderabad.
42. G.P. Chapman and Sudhir Wanmali, (1972) "Urban-Rural Relationships in India : A Macro-Scale Approach Using Population Potentials", *Geoforum*, 12, pp. 19-44.
43. S. Manzoor Alam and Waheeduddin, Khan, (1972) *Metropolitan Hyderabad and Its Region : A Strategy for Development*, Asia Publishing House, Bombay.
44. J. Holmes, F.B. Williams and L.A. Brown, (1972) "Facility Location under a Maximum Travel Restriction : An Example Using Daycare Facilities", *Geographical Analysis*, 4, pp. 258-66.
45. S. Wanmali and Avijit Ghosh, (1975) "Pattern of Distribution of Consumer Goods in Rural India", *Management and Labour Studies*, 1, pp. 79-94.
46. K.N. Gopi, (1978) *Process of Urban Fringe Development : A Model*, Concept Publishing Company, New Delhi.
47. C.D. Deshpande, B. Arunachalam and L.S. Bhat, (1979) "South Kolaba — A Study in Settlement Hierarchy and Rural-Urban Interaction", *Mimeographed Report of an I.C.S.S.R. Sponsored Project* (New Delhi).
48. V.L.S. Prakasa Rao, and V.K. Tewari, (1980) *The Structure of an Indian Metropolis : Bangalore — An Urban Survey*, Allied Publishers, New Delhi.
49. M. Mortimore, (1981) "An Alternative Opportunity Framework for Population Mobility, Illustrated from Nigeria" in *Population Distribution and Re-distribution in Africa* in J.I. Clarke and L.A. Kosinski (eds.), Heinemann, London.
50. B. Banerjee, (1987) *Rural to Urban Migration and the Urban Labour Market: A Case Study of Delhi*, Himalaya Publishing House, Delhi, pp. 1-279.

2

SPATIAL AND TEMPORAL FLOW OF COMMODITIES

INTRODUCTION

THIS CHAPTER is devoted to the study of flow of commodities in spatial and temporal terms. Places depend upon each other for several commodities. Meerut is known for its agricultural as well as industrial advancement. As a result, field of interaction has expanded well beyond regional limits. Production of food crops, vegetables, fruits, industrial and fodder crops has led to the growth of industrial and commercial activities in the Meerut city. Due to modernisation of the production technology, huge quantity of agricultural as well as industrial surplus is generated. Meerut city acts as a nodal centre through which these surpluses are supplied out to different parts of the country. In the course of development, Meerut has also experienced greater dependency to other areas for raw materials, construction goods, defence goods and other items. Extent of flow over time and space explains the limits and gravity of space relations in terms of the amount of surplus and deficit. It is possible to recognise functional (amount) and areal gaps through flow differentials and to measure the income inequalities. This chapter has been broadly sub-divided into four sections, viz.:

(a) Inflow of commodities,
(b) Outflow of commodities,

(c) Nature of surplus and deficit commodities, and
(d) Functional gaps and income inequalities.

This chapter is based on the data obtained from the offices of *Mandi Parishad*, Octroi Records, Assistant Director of Industries, Meerut, Feeder Balancing Dairy, Partapur, and the amount of goods booking through railways and highways. Railway and textile records were available for the period of 1975-1984, *Mandi* and Octroi for the period of 1981-1984 and Feeder Balancing Dairy, Partapur, for the period of 1973-1984. While data on arrival of commodities are well recorded, data on outflow of commodities are not recorded so well. Railway records for arrival and outflow are available on the same pattern whereas highway records for arrival and outflow are not on the same pattern. While arrival data in case of highways are recorded for all commodities, outflow data are recorded only for selected commodities (taxable or restricted commodities). Both statistical as well as cartographic methods have been used to elaborate the results. Details about the methods used and maps constructed are given along with the text.

Inflow of Commodities

Meerut recorded an annual inflow of 2.52 million tonnes of commodities during 1984. The share of arrival through highways was about 72 (71.89) per cent and through railways 28 (28.11) per cent. Since both the ways are significant in terms of arrival, a detailed analysis of the types of commodity arrival, their amount and place of origins is needed. Separate account of inflow of commodities through railways and highways are given as under:

Arrival Through Railways

Meerut recorded an arrival of about seven lakh tonnes (Table 2.1) of heavy goods and about 9,000 tonnes of light goods[1] through railways during 1984. Fertilizer is the chief arrival under heavy goods which shared about 30 (30.42)[2] per cent of the weight. It is followed by defence goods (17.57%) and coal (13.98%). Firewood and petroleum both accounted for about 12 (12.48%) separately. Remaining commodities were less significant in terms of arrival as they together shared about nine (9.53) per cent of the weight. Cement is one of the chief city demands which accounted for about four (3.54) per cent of the total arrival. Rest of the arrival is grouped under miscellaneous goods that in addition to several low weight items also includes timber, oilcake and railway material.

TABLE 2.1: Meerut City — Spatial Trends of Arrivals Through Railways — 1984

(Weight : In Tonnes)

Commodities	*States*	*Percentage share of constituent States*	*Arrival*	*Percentage to the total arrival*
Heavy goods				
Petroleum	Gujarat	72.00	87,229	12.48
	U.P	28.00		
Coal	Bihar	60.00	97,700	13.98
	West Bengal	40.00		
Firewood	U.P	100.00	87,232	12.48
Logs	Orissa	64.97	1,276	0.18
	M.P.	35.03		
Timber	Bihar	84.99	3,258	0.47
	West Bengal	10.01		
	Assam	5.00		
Oilcake and gram husk	U.P.	65.00	2,140	0.31
	M.P.	35.00		
Cement	M.P.	85.00	24,737	3.54
	Rajasthan	15.00		
Fertilizer	Gujarat	52.00	2,12,568	30.42
	Maharashtra	35.00		
	Rajasthan	8.00		
	U.P.	5.00		
Railway material	Delhi	80.00	460	0.06
	U.P.	20.00		
Miscellaneous goods	West Bengal	67.50	59,496	8.51
	U.P.	17.50		
	Maharashtra	11.25		
	J&K	3.75		
Defence goods	Delhi	85.00	1,22,763	17.57
	Punjab	7.00		
	U.P.	5.00		
	Maharashtra	3.00		
Total			6,98,859	100.00

Contd.

TABLE 2.1 *(Contd.)*

Commodities	*States*	*Percentage share of constituent States*	*Arrival*	*Percentage to the total arrival*
Light goods				
Printed books	U.P.	60.00	3,072	33.93
	Tamil Nadu	25.00		
	Kerala	15.00		
Medicines	Maharashtra	50.00	1,168	12.90
	Gujarat	34.93		
	Delhi	15.07		
Sports goods	Punjab	79.94	982	10.84
	West Bengal	20.06		
Handloom cloth	Maharashtra	59.17	1,690	18.66
	Tamil Nadu	40.83		
Aluminium	Delhi	100.00	384	4.24
Handloom machine	Gujarat	70.31	384	4.24
	Punjab	29.69		
Defence goods	West Bengal	32.00	1,375	15.19
	Delhi	44.00		
	Punjab	18.62		
	U.P.	5.38		
Total			9,055	100.00

Spatial flow of commodities reveal that fertilizer had an annual inflow 2,12,000 tonnes. Of the total fertilizer arrival, Gujarat accounted for 52 per cent followed by Maharashtra 35 per cent. Besides, fertilizer was also received from the States of Rajasthan (8%) and Uttar Pradesh (5%). Defence goods are next in order of importance and recorded an annual inflow of about one (1.2) lakh tonnes. Of the total arrival of defence goods, Delhi accounted for 85 per cent and States of Punjab, Uttar Pradesh and Maharashtra for about 15 per cent. Coal ranked third in arrival. About 98 thousand tonnes of coal was received in the city during 1984. Of the total arrival, Bihar accounted for about 60 per cent and West Bengal for the remaining 40 per cent. Petroleum accounted for about 87 thousand tonnes of arrival during 1984. The major receipt of petroleum, i.e., about 72 per cent, was from the State of Gujarat and a minor quantity of receipt, i.e., about 28 per cent, was from Uttar Pradesh. Arrival of firewood in the city accounted for 87 thousand tonnes and was entirely from different parts of Uttar Pradesh. Cement inflow was about 5 thousand tonnes. Madhya Pradesh (85%) is the chief supplier of cement to the city. A small quantity of cement was also received from the State of Rajasthan (15%). Miscellaneous goods accounted for about 60,000 tonnes. Such goods are supplied from the States of West Bengal (67.50%), Uttar Pradesh (17.50%), Maharashtra (11.25%) and Jammu & Kashmir (3.75%). Apart from these commodities Meerut also recorded inflow of timber from Bihar, West Bengal and Assam; logs from the States of Orissa and Madhya Pradesh; oilcake and gram husk from Uttar Pradesh and Madhya Pradesh and railway material from the Union Territory of Delhi and the State of Uttar Pradesh.

Arrival of light goods registered maximum inflow of printed books (33.93 per cent of the total) which is followed by the handloom cloth (18.66%) and defence goods (15.19%). Medicines and sports goods are also significant in terms of arrival as they accounted for 13 (12.90) and 11 (10.84) per cent, respectively. Arrival of aluminium and handloom machines was the minimum, i.e., 4.24 per cent each. In terms of weight, Meerut recorded an arrival of 3,000 tonnes of books during 1984. Arrival of printed books was mostly from Uttar Pradesh (60%) followed by Tamil Nadu and Kerala. Handloom cloth registered an inflow of 1,700 tonnes of weight. Arrival of cloth was mostly from Maharashtra (57.17%) and Tamil Nadu (40.83%). Defence goods weighing 1,375 tonnes were supplied from Delhi, West Bengal, Punjab and Uttar Pradesh. About 1,168 tonnes of medicines were received from Maharashtra (50%), Gujarat (34.93%) and Delhi (15.07%). Sports goods worth about 1,000 tonnes were received from Punjab (79.94%) and West Bengal (20.06%). Aluminium goods and handloom machines recorded inflow of 384 tonnes each. Supply of aluminium is mostly from Delhi whereas of handloom machines from Gujarat and Punjab.

It is, thus, clear that Meerut city depends on different parts of the country for a variety of goods and commodities which are not produced in the city or in the region. Railways act as the chief link between areas of supply and demand. While it holds true that most of the arrivals are not produced in the city or in the region, a few commodities, tend present a departure from this generalisation. Commodities, like, sports goods, printed books and handloom cloth, which are surplus in the production, have been recorded as arrival. Such an arrival causes contradiction to the understanding of surplus-deficit realms.

Further probe into the problem revealed that production, being a complex phenomenon, may or may not depend upon its region entirely. It may prefer to depend upon other centres for quality production. Besides quality, variety of the products may also be the choice depending upon the purchasing power of the people and market competition. Meerut, besides being a surplus centre of certain goods and commodities, is at the same time a place of demand too for certain goods and commodities.

The spatial distribution of the inflow (Fig. 2.1) of commodities reveals two distinct patterns. First, the arrival of commodities tends to decrease with the distance. While light goods tend to arrive at Meerut from all parts of the country, heavy goods arrive here mostly from the northern parts of the country particularly from neighbouring districts. Second, the arrival of commodities tends to specialise with the distance. Goods and commodities, in many numbers and large quantities, arrive here from neighbouring areas whereas limited number of commodities in limited quantities arrive from distant places.

Temporal trend (Table 2.2) of receipts recorded increasing pattern both in case of light as well as in heavy goods. Light goods registered a total inflow of about four thousand tonnes in 1975 which increased to about nine thousand (9,055) tonnes by 1984, registering over 100 per cent rise in the arrival. Similarly, Meerut city recorded about 5 lakh tonnes of heavy goods in 1975, which increased to about 7 lakh tonnes by 1984. Thus, over a period of nine years, heavy goods registered a rise of about 40 per cent. The overall inflow of all types of goods (light + heavy goods) was little more than 5 lakh tonnes in 1975 which registered the level of over 7 lakh tonnes at the end of 1984. Results reveal major variations in the annual inflow patterns of heavy goods. The deviation is, however, much less in case of light goods. The analysis of monthly (Fig. 2.2) trend of arrival revealed minimum receipt of light goods in the month of January (583 tonnes) and maximum in the month of December (911 tonnes). These goods continue to increase or remain at the same level during the months of April to December. From January to March arrival of these goods presents uneven trend. On the contrary, monthly trend of heavy goods recorded minimum arrival in the month of September and maximum in the month of December. These goods present a bi-modal situation in the arrival trend. Quantity of arrival falls sharply between March to September and rises steeply between October to December. The trend of monthly arrival is uneven presenting sharp monthly variations. The overall trend of inflow seems to be affected by the pattern of heavy goods. It is marked with a minimum of about five (4.80) per cent in the month of September to a maximum of 12 (12.30) per cent in the month of December.

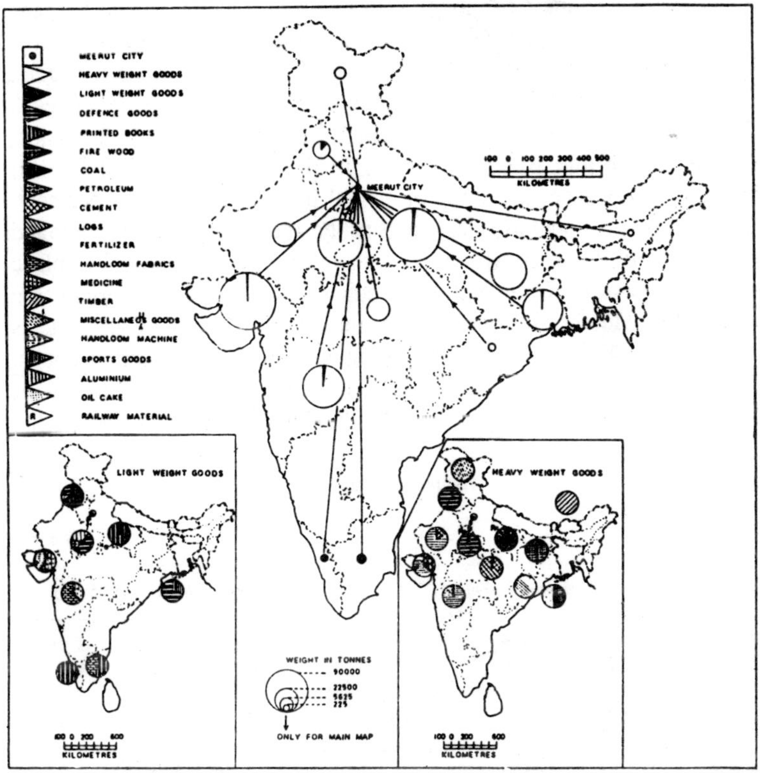

FIG. 2.1: Meerut city — Inflow of commodities through railways — 1984.

Arrival Through Highways

Octroi: Octroi (city checkposts at each entry point) records tender useful information about the arrival of commodities in the city. About 16 (16.35) lakh tonnes (Table 2.3) of commodities were reported to have arrived in Meerut city during 1984. Of the total arrival about 42 per cent related to the construction goods, 30 per cent fuel, 19 per cent food items, five per cent to each cloth and metal and remaining less than one per cent to leather. Commodities registering extremely low weight were put together and were explained in the groups of other commodities. The relative importance of routes is revealed through respective shares in the total arrival. Delhi road is the most important route which accounted for about 47 per cent of the total arrival. Next in order of importance are Garh and Hapur routes with their respective shares as 13 (12.57) and 11 (11.41) per cent. Muzaffarnagar, Baraut, Baghpat and Mawana routes are in the third order of importance, each accounting for about eight per cent of the total arrival. Sardhana, Parikshhatgarh and Bhola routes are in the fourth (the lowest level) order of importance, the share of arrival being less than one per cent. Growing urban population has caused the increasing housing demand in the city. Building construction is, therefore, gaining the prime place in urban persuits. Construction goods such as sand and bricks are supplied to Meerut city mostly from neighbouring areas. Garh road is the most significant link accomplishing demand of construction goods. This route accounted for 26 (26.42) per cent of the total arrival of construction goods during 1984. Hapur and Baraut routes are at the

TABLE 2.2: Meerut City — Temporal Trends of Commodity Arrivals Through Railways

(*Weight : In Tonnes*)

	Monthly trends (1984)			*Annual trends (1975-84)*			
Months	*Light goods*	*Heavy goods*	*Share of monthly arrival*	*Years*	*Light goods*	*Heavy goods*	*Growth rate*
January	583	63,863	9.10	1975	3,989	4,99,623	—
February	585	68,252	9.72	1976	5,447	7,26,787	45.40
March	596	78,728	11.20	1977	7,698	8,32,599	14.76
April	739	48,653	6.98	1978	7,750	6,27,389	−24.41
May	694	48,757	6.99	1979	8,359	4,52,927	−27.37
June	862	54,353	7.80	1980	7,844	6,84,033	49.99
July	819	51,184	7.35	1981	9,159	6,87,026	0.62
August	831	47,182	6.78	1982	6,914	6,75,840	1.93
September	795	33,178	4.80	1983	7,144	8,20,436	21.21
October	880	58,083	8.33	1984	9,055	6,98,859	−14.46
November	760	60,467	8.65				
December	911	86,159	12.30				

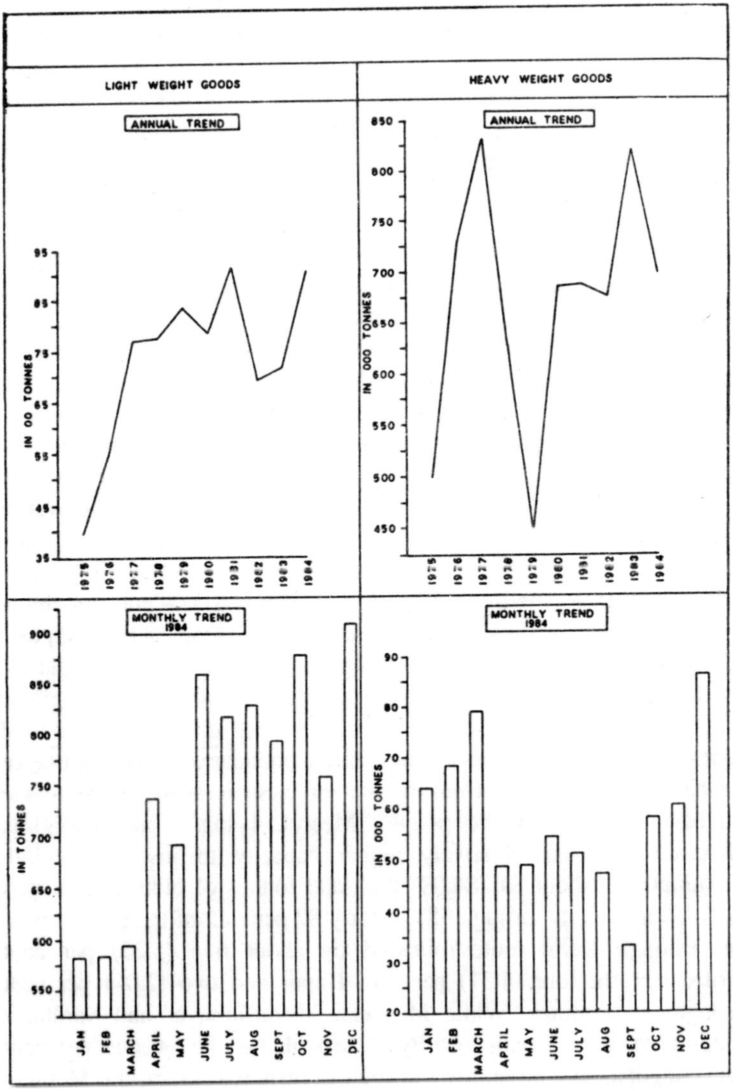

FIG. 2.2: Meerut city — Inflow trends of commodities through railways.

same level of importance with respective arrivals as 19 and 18 per cent. While Baghpat, Mawana and Muzaffarnagar routes are at the third level of arrival, Delhi, Parikshhatgarh, Sardhana and Bhola routes recorded the fourth level being at minimum importance in terms of arrival. Most of the construction goods except bricks and sand arrive here from far off region. Riverside areas in the district are the main source of the supply of sand to the city. Bricks are supplied from the bricks fields which form a typical landscape beyond the city limits. Large scale construction works are being carried out in the city along Hapur, Garh, Muzaffarnagar and Sardhana routes. The inflow of construction goods is maximum through Garh route (26.42%) followed by Hapur (19.37%), Baraut (18%) and Baghpat (17.22%) routes. Fuel is one of the essential requirements of the household and transport sectors. In terms of weight, fuel ranks second accounting for about 30 per cent of the total city arrivals. Most of the fuel supply to the city is through Delhi route. About 99 per cent of the fuel arrival is through this route. A small quantity (below 1 per cent) of fuel arrival is also recorded through Hapur and Garh routes. Food items arrive in the city from all directions. The quantity of arrival through these routes varies significantly. Delhi, Muzaffarnagar and Mawana routes together accounted for over 85 per cent of the food arrival. Nearly 13 per cent of the food arrival is through Garh and Hapur routes and remaining about two per cent through other routes. From these facts, it appears that Meerut attracts surplus food items from distant regional places. The fact that national highways carry over a larger share of food grains than the State and district highways, it seems that some amount of food items arrive here from across the district boundaries. Cloth is one of the basic human needs. Consumption of cloth is proportional to the purchasing power of the people. Meerut, being economically an advanced area, has relatively higher capacity to invest for cloth consumption. About 84 .(84.08) thousand tonnes of cloth arrival was recorded in the city during 1984. Nearly 87 per cent of the cloth arrival was through Delhi route followed by about five (5.25) per cent Parikshhatgarh, four (4.37) per cent Hapur and two (2.04) per cent through Garh routes. While Mawana, Muzaffarnagar and Sardhana routes recorded a small quantity of arrival, i.e., below one per cent each, Baghpat, Baraut and Bhola routes registered no arrival. Metal is one of the chief trading commodities of the Meerut city. Meerut recorded an annual inflow of metal worth about 83 (83.22) thousand tonnes during 1984. Arrival of metal goods is mostly through Delhi and Hapur routes. These two routes together account for more than 98 per cent of the total arrival. Baraut, Parikshhatgarh and Sardhana

TABLE 2.3: Meerut City — Inflow of Commodities Through Highways During — 1984 (Based on Octroi Records)

(Weight : In Tonnes)

Routes	Commodities																						
	Construction goods			Fuel			Food items			Cloth			Metal			Leather			Other commodities			Total arrival	% share
	A	P	R	A	P	R	A	P	R	A	P	R	A	P	R	A	P	R	A	P	R		
1	2	3	4	5	6	7	8	9	10	11	12	13	14	15	16	17	18	19	20	21	22	23	24
Delhi	24,022.35	3.52	7	472775.30	98.9	1	138288.5	45.4	1	73516.0	87.4	1	52429.6	63.0	1	206.9	28.9	2	1826.3	73.6	1	763064.9	46.7
Muzaffarnagar	38,080.90	5.58	6	27.31	0.0	4	95059.3	31.2	2	142.9	0.2	6	10.0	0.0	7	1.4	0.2	7	0.6	0.0	3	133322.4	8.1
Mawana	60,124.23	8.81	5	19.56	.0	6	27728.75	9.1	3	487.6	0.6	5	14.1	0.0	6	-	-	-	-	-	-	88374.3	5.4
Parikshhatgarh	3,002.80	0.44	8	10.05	.0	10	152.3	0.0	10	4413.9	5.2	2	-	-	-	413.79	57.8	1	-	-	-	7992.9	0.5
Garhmukteswar	1,80,304.14	26.42	1	860.22	0.2	3	21804.6	7.2	4	1715.1	2.0	4	174.3	0.2	4	82.7	11.6	3	655.1	26.4	2	205596.7	12.6
Hapur	1,32,191.18	19.37	2	4157.84	0.9	2	17419.3	5.7	5	3674.1	4.4	3	29153.44	35.03	2	6.87	0.9	4	-	-	-	186602.8	11.4
Sardhana	2,388.58	0.35	9	11.00	.0	9	609.1	0.2	8	126.1	0.1	7	-	-	-	-	-	-	-	-	-	3134.8	0.2
Baghpat	1,17,518.40	17.22	4	19.68	.0	5	1857.6	0.6	6	-	-	-	306.58	1.6	3	-	-	-	-	-	-	120702.3	7.4
Baraut	1,22,841.58	18.00	3	13.80	.0	8	1248.6	0.4	7	-	-	-	-	-	-	2.7	0.4	5	-	-	-	124106.7	7.6
Bhola	1,979.12	0.29	10	15.24	0.0	7	365.4	0.1	9	-	-		133.1	0.16	5	1.6	0.2	6	-	-	-	2494.54	0.15
Total	682453.28			477910.0			304533.42			84076.42			83221.7			715.9			2481.9			1635392.37	100.0
Percentage to total city inflow	41.73			29.22			18.62			5.14			5.10			0.04			100.0			100.00	

Note: A = stands for total arrival in tonnes; P = stands for percentage share, and R = stands for rank.

routes did not register any arrival, whereas other routes registered nominal inflow. Leather is recorded as one of the arrivals in the city. The total leather arrival in the city was about 700 tonnes during 1984. Arrival of leather is mostly through Parikshhatgarh (57.79%), Delhi (28.90%) and Garhmukteswar (11.56%) routes. While Sardhana, Mawana and Baghpat routes did not register any quantity of leather arrival; Baraut, Bhola, Muzaffarnagar and Hapur routes registered a very nominal arrival. There are several other commodities that have registered small quantity of annual arrival. As discussion about each one of them is not so meaningful, all these commodities have been discussed under other commodities. About two and half (2.48) thousand tonnes of arrival was recorded as inflow of other commodities. Delhi, Garhmukteswar and Muzaffarnagar routes recorded most of arrival of such commodities.

Spatial trend to arrival (Fig. 2.3) reveals that construction goods are the main items of arrival from Muzaffarnagar, Baraut, Bhola, Baghpat, Garhmukteswar and Mawana routes. As such construction goods arrive from almost all the directions but they specialise more towards western and northern routes. Delhi route, the major direction of arrival of most of the commodities, is the sole supplier of petroleum, fuel, a large quantity of food items, cloth, metal, leather and other commodities. With exception to construction and leather goods, Delhi route ranked first in all items of arrival. While Muzaffarnagar route recorded a relatively higher amount of the inflow of food items, Parikshhatgarh route recorded a higher proportion of cloth and leather arrival to the city. Alike other routes, Hapur route recorded relatively higher share of metal and fuel arrival. It is, thus, obvious that Meerut city has directional specialisation for the arrival of various commodities. Delhi route is the most important artery of the arrival of commodities to the Meerut city. Garhmukteswar route has specialised in the inflow of construction goods, Parikshhatgarh route in leather goods and Muzaffarnagar route in food items.

Temporal trend of arrival (Fig. 2.5) recorded regular increase in the quantity of arrival ever since 1981. Meerut registered about 6 lakh tonnes of commodity arrival during 1981. Inflow rose to the level of 1.63 million tonnes within a short span of three years (1984). The annual growth rate during 1981-82 was about 80 per cent, 1982-83 about 25 per cent and 1983-84 about 20 per cent. Thus, growth rate registered a down trend in successive years. It is, however, important to note that annual growth rate was never below 19 per cent. The monthly behaviour of arrival was influenced by the seasonal factors. For instance, there was a decline in the quantity of arrival during April to August, an increase during September to December and almost even arrival during January to March. While August registered the minimum quantity of arrival, i.e., 91 thousand tonnes, December registered the maximum arrival of about 2 (1.70) lakh tonnes during 1984.

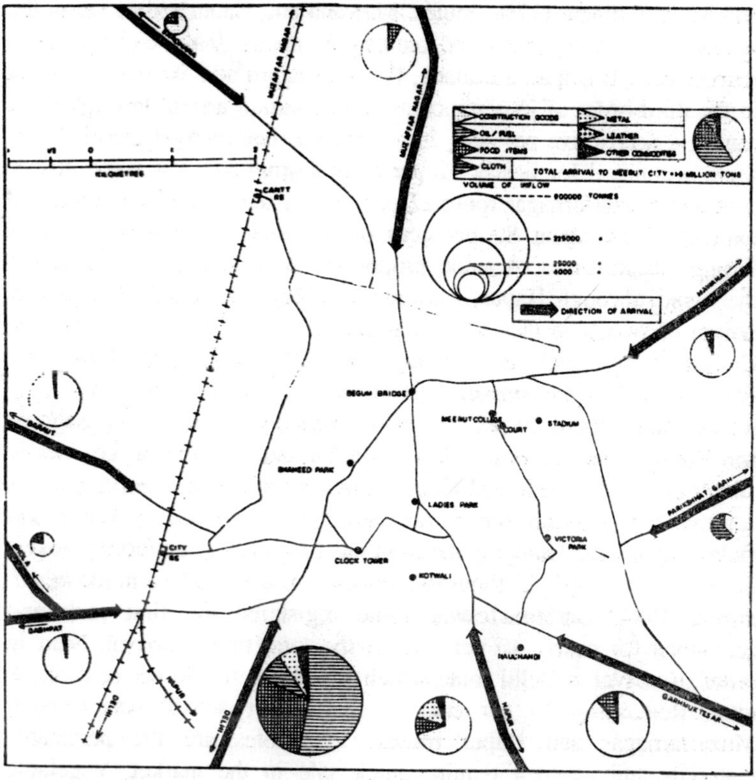

FIG. 2.3: Meerut city — Inflow of goods and commodities (based on octroi records) — 1984.

Mandi: *Mandi* acts as a major centre of exchange of commodities. Meerut has a well-developed *mandi* system through which it maintains flow relations (inflow as well outflow) of commodities with various places in India. Table 2.4 reveals that about 1½ lakh tonnes (1.54) of commodities were recorded as arrival and about five lakh (5.23) tonnes as outflow from Meerut *mandi*. Of the total arrival about 42 (41.66) per cent was of foodgrains, 24 (24.07) per cent sweets, (Gur, Khandsari etc.), 14 (14.03) per cent timber, 10 (10.01) per cent vegetables, five (5.38) per cent livestock feed, four (3.53) per cent fruits and one (1.32) per cent other commodities. In terms of relative strength, Garhmukteswar route is the most important route accounting for about 28 per cent of the total *mandi* arrival. It is followed by Delhi (23%) and Muzaffarnagar (17%) routes. Mawana and Hapur routes are in the second order of importance accounting for about 10 and eight per cent, respectively. Baghpat, Sardhana, Parikshhatgarh and Baraut routes are in the third order of importance as annual *mandi* arrival through these routes is nearly five per cent. In terms of arrival of food grains, Delhi route accounted for nearly 32 per cent, Garhmukteswar route 26 per cent and Muzaffarnagar route 22 per cent of the total (64.25 thousand tonnes). Thus, about 80 per cent of the total foodgrains arrival is through these routes. A small proportion of foodgrain also arrives in the *mandi* through Hapur, Mawana and Baghpat routes. The annual arrival of sweets in the *mandi* was about 37 (37.13) thousand tonnes. Garhmukteswar route accounted for 22 (22.39) per cent of the total sweet arrival in the *mandi*. It is closely followed by Mawana route which shared about 20 per cent of the total arrival. Sardhana, Baghpat and Baraut routes recorded almost similar share of arrival, i.e., nearly 10 per cent each. Delhi and Hapur routes are the least important in this context. It is possibly due to the pull effect exercised by Hapur and Delhi *gur mandi*. Timber is the third most important commodity arrival in the *mandi*. About 22 thousand tonnes of timber arrived in the *mandi* during 1984. Garhmukteswar route registered the first rank and accounted for nearly 40 per cent of the total timber arrival. Next in order of arrival is Delhi route which accounted for 30 per cent of the total. Remaining 30 per cent of the timber arrival was through Muzaffarnagar and Hapur routes. Vegetables are the perishable materials and as such require quick sale in the market. Vegetable arrival in the *mandi* was about 15.44 thousand tonnes during 1984. Garhmukteswar route (20.51%) ranked first followed closely by Muzaffarnagar route (20 %). Parikshhatgarh, Delhi and Baghpat routes are next in order of importance accounting for about 14 (13.56), 10 (10.26) and 10 (9.74) per cent, respectively. Hapur, Mawana, Sardhana and Baraut routes are the least important in terms of vegetable arrival to Meerut *mandi*. Vegetable farming is widely practised near Hapur but its supply to Meerut is meagre because most of it is directed towards Delhi. The functions of nearest neighbour and distance decay does not operate because of the great pull effect of Delhi vegetable market. Thus, produce moves straight to Delhi by passing nearest market Meerut. Apart from vegetables, livestock feed is also supplied to the *mandi* in a huge quantity. An arrival of about 8.30 thousand tonnes of livestock feed was recorded in the *mandi* during the period under study. It moved mainly through Garh, Mawana and Muzaffarnagar routes accounting for 16, 13 and 10 per cent, respectively. Fruit farming is one of the increasing horticultural activities of rural areas in the district. Meerut is, therefore, known for its fruit collection, production and supply. Meerut *mandi* recorded an arrival of about 5 (5.44) thousand tonnes of fruit during 1984. The maximum flow of fruits was through Delhi route (60%), followed by Muzaffarnagar, Garh and Hapur routes. Delhi, being a major fruit market, meets most of the winter season demand of the city. Meerut becomes surplus in fruits during summer season. Fruits' arrival during summer season is from neighbouring areas, i.e., mostly through Hapur, Garh and Muzaffarnagar routes. Besides these commodities, about 2.03 thousand tonnes of other commodities were also received under the heads of fibre products, spices, animal products and smokes.

Spatial structure (Fig. 2.4) of *mandi* arrival reveals that foodgrains largely come from the national highways, like, Delhi, Muzaffarnagar, Hapur and Garhmukteswar. These routes have recorded more than the average proportion of foodgrain arrival in the *mandi*. It is interesting to note that other routes have either no inflow or have much less than the average. On the contrary, these routes specialise in the inflow of sweets. Sweets accounted for two-thirds to three-fourths of the total inflow from these routes. Arrival of timber is mostly from the southern routes, i.e., about 80 per cent of the total inflow. Livestock feed arrived mostly from Mawana and Garhmukteswar routes whereas fruits from Delhi, Muzaffarnagar and Garhmukteswar routes. Vegetable is supplied to the city from all directions. It is, therefore, possible to generalize that vegetable and sweets have a flow from all directions whereas fruits, livestock feed, timber and foodgrains tend to specialise along certain directions, like, Delhi, Muzaffarnagar and Garhmukteswar routes. While Delhi route ranked first in overall arrival to the city, Garhmukteswar route ranked first in the *mandi* arrivals.

TABLE 2.4: Meerut City — Mandi Arrivals Through Highways During 1984

(Weight : In Tonnes)

Routes	Commodities																						
	Foodgrains			*Sweets*			*Timber*			*Livestock feed*			*Vegetables*			*Fruits*			*Other commodities*			*Total arri-val*	*% share*
	A	P	R	A	P	R	A	P	R	A	P	R	A	P	R	A	P	R	A	P	R		
1	2	3	4	5	6	7	8	9	10	11	12	13	14	15	16	17	18	19	20	21	22	23	24
Delhi	20,561.00	32.00	1	1855.3	5.0	8	6492.8	30.0	2	-	-	-	1584.0	10.3	4	3264.8	60.0	1	1229.4	60.7	1	34988.3	22.7
Muzaffarnagar	14,135.70	22.00	3	3013.8	8.1	6	4328.5	20.0	3	830.0	10.0	3	3087.5	20.0	2	1088.2	20.0	2	185.1	9.1	4	26668.8	17.3
Mawana	4,124.30	6.42	5	7425.3	20.0	2	-	-	-	2489.8	30.0	2	993.8	6.4	7	-	-	-	-	-	-	15033.2	9.7
Parikshhatgarh	-	-	-	2824.80	7.6	7	-	-	-	-	-	-	2093.6	13.6	3	-	-	-	-	-	-	4918.4	3.2
Garhmukteswar	16,705.90	26.00	2	8313.2	22.4	1	8657.0	40.0	1	4979.7	60.0	1	3166.8	20.5	1	547.2	10.1	3	329.7	16.3	2	42699.5	27.7
Hapur	6,156.30	9.58	4	1743.5	4.7	9	2164.2	10.0	4	-	-	-	1464.4	9.5	6	541.0	9.9	4	282.6	13.9	3	12352.0	8.0
Sardhana	-	-	-	4411.6	11.9	3	-	-	-	-	-	-	859.8	5.6	8	-	-	-	-	-	-	5271.4	3.4
Baghpat	2,570.00	4.00	6	3712.7	10.0	5	-	-	-	-	-	-	1503.4	9.7	5	-	-	-	-	-	-	7786.1	5.0
Baraut	-	-	-	3825.5	10.3	4	-	-	-	-	-	-	683.8	4.4	9	-	-	-	-	-	-	4509.3	2.9
Total	64,253.20			37136.7			21642.5			8299.5			15437.1			5441.2			2026.8			154227.0	100.0
Percentage to total arrival in the city	41.7			24.1			14.0			5.4			10.0			3.5			1.3			100.0	

Note: A = stands for arrival in tonnes; P = stands for percentage share and R = stands for rank.

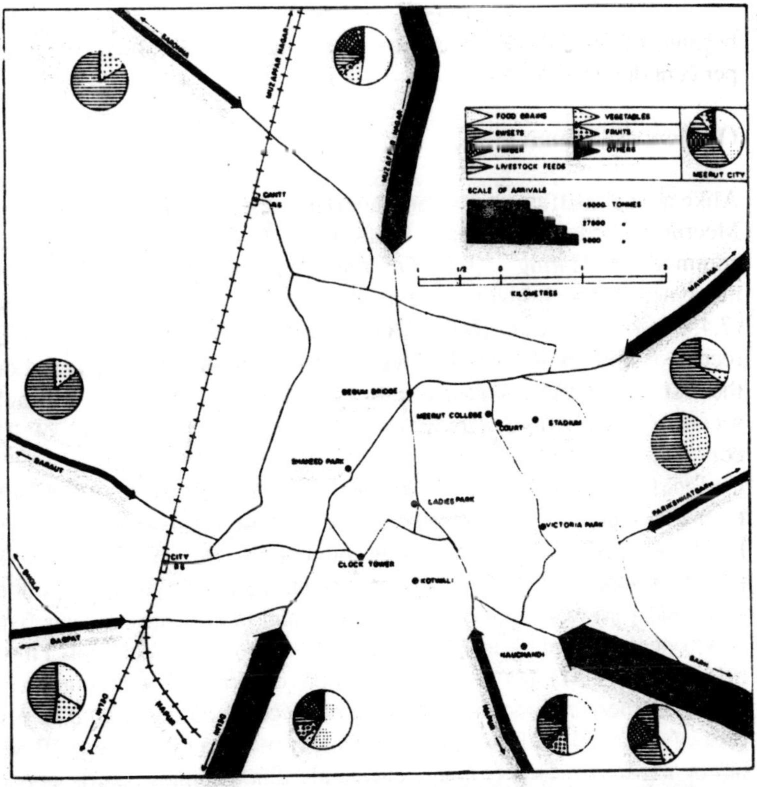

FIG. 2.4: Meerut city — Structure of *mandi* arrival — 1984 ,

Temporal trend (Fig. 2.5) explains that January and June months are the two peak periods of *mandi* arrival. These months account for more than 15 per cent of the annual arrival. Months of July and September registered the minimum monthly arrival. Temporal trend is characterised by the increasing behaviour of arrival from October to January and decreasing from February to April with steep rise in May and June whereas fall in July and September. There are marked monthly variations in the quantity of arrival, i.e., from 5 thousand to 25 thousand tonnes. With the result coefficient of variation is quite high. The annual trend of arrival (Table 2.5) explains gradual increase in the quantity of arrival. However, a decline of 3.57 per cent was observed during 1983. This is possibly due to the poor harvests in the period because of the drought conditions. Subsequently, arrival rose by 10.87 per cent during 1983-84.

Outflow of Commodities

Alike arrival, outflow is also an important aspect of interaction studies. Meerut recorded an annual outflow of 6 (6.45) lakh tonnes of commodities during 1984. Of the total outflow, proportion of movement through highways and railways was 93 (92.55) and seven (7.45) per cent, respectively. Due to non-availability of the official record of outflow through highways, interpretation is largely based on the data of outflow through railways. Besides explanation through secondary data, field survey was also carried out in case of 15 commodities supposed to be in surplus production from the city.[3]

Field survey revealed an annual outflow of about 6 (5.97) lakh tonnes of commodities through highways. Of the total outflow through highways, agricultural products accounted for 60 (59.59) per cent, rubber products 12 (12.22) per cent, sugar 10 (9.55) per cent, cotton cloth and yarn eight (8.47) per cent, milk and milk-made products five (5.45) per cent and agricultural implements two (1.53) per cent. Thus, nearly 97 (96.81) per cent of the total outflow through highways related to these commodities. Among other commodities (i.e., 3.19 per cent of the total outflow through highways) registering low outflow, liquor registered 5,400 tonnes, chemicals 3,000 tonnes, sports goods 1,854 (1,854.63) tonnes, *gobar* gas equipments 1,825 tonnes and pharmaceutical sugar cube 1,800 tonnes per annum. Transformers accounted for 1,500 tonnes, surgical goods 600 tonnes, solder-wire 420 tonnes and cutlery items 87 (87.03) tonnes.

Outflow of Commodities through Railways

Railways are the chief modes of transport for the supply and receipt of the goods in the country. Meerut is surplus in several commodities, as such, some of these commodities are supplied to different parts of the country through railways. Meerut city registered (Table 2.6) supply of 19 commodities weighing about 48 thousand tonnes during 1984. According to the weight, these commodities are grouped as light and heavy, their respective weights being 6 (6.5) and 42 (41.5) thousand tonnes. About 14 commodities were supplied through railway bookings (heavy goods) and remaining five commodities through railway parcels (light goods).

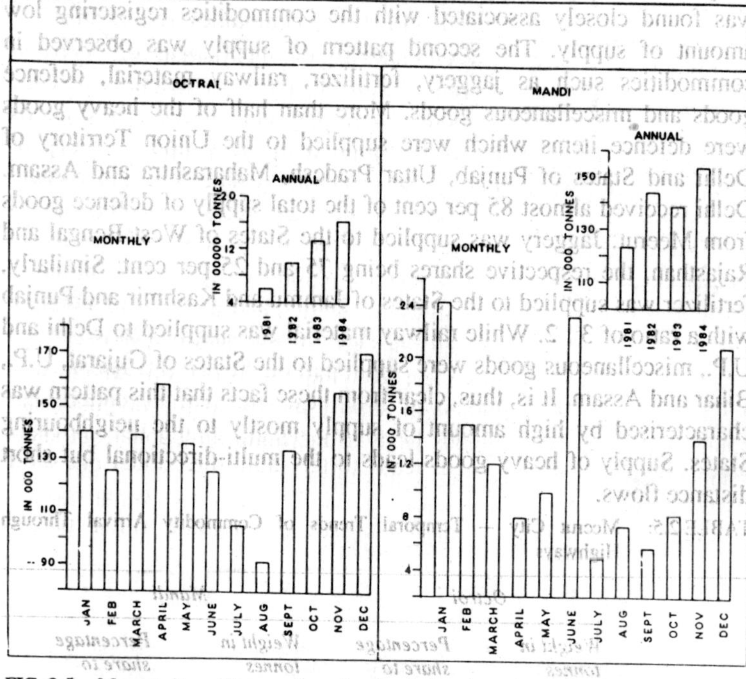

FIG. 2.5: Meerut city – Temporal trends of arrival – (octroi and *mandi*).

Heavy goods form the major share of supply through railways. Of the total outward supply of these goods, about 51 per cent were defence goods followed by jaggery 25 per cent. Khandsari molasses, crushed bone, wheat, khandsari, railway material and miscellaneous goods together accounted for nearly 20 per cent, the respective shares ranged between one to six per cent. Commodities like fertilizer, waste cotton, *gawar*, floor, potato and matting accounted for about four per cent of the total outflow under this category. Based on the spatial pattern of outflow, heavy goods may be sub-divided into two categories. The first category includes those commodities which are en *bloc* supplied to a single State whereas the second category includes those commodities which are supplied to many States. Of the 14 commodities supplied under heavy goods, nine belong to the first category and remaining five to the second. Commodities such as potato, wheat, waste cotton, khandsari molasses, *gawar*, matting and floor are enlisted in the first category of goods. Among these goods, first four commodities were solely supplied to West Bengal, *gawar* to

Punjab, matting to Haryana and floor to Uttar Pradesh. This pattern was found closely associated with the commodities registering low amount of supply. The second pattern of supply was observed in commodities such as jaggery, fertilizer, railway material, defence goods and miscellaneous goods. More than half of the heavy goods were defence items which were supplied to the Union Territory of Delhi and States of Punjab, Uttar Pradesh, Maharashtra and Assam. Delhi received almost 85 per cent of the total supply of defence goods from Meerut. Jaggery was supplied to the States of West Bengal and Rajasthan, the respective shares being 75 and 25 per cent. Similarly, fertilizer was supplied to the States of Jammu and Kashmir and Punjab with a ratio of 3 : 2. While railway material was supplied to Delhi and U.P., miscellaneous goods were supplied to the States of Gujarat, U.P., Bihar and Assam. It is, thus, clear from these facts that this pattern was characterised by high amount of supply mostly to the neighbouring States. Supply of heavy goods leads to the multi-directional but short distance flows.

TABLE 2.5: Meerut City — Temporal Trends of Commodity Arrival Through Highways

	Octroi		*Mandi*	
	Weight in tonnes	*Percentage share to total arrival*	*Weight in tonnes*	*Percentage share to total arrival*
Monthly trends (1984)				
January	1,40,115	8.57	24,230	15.71
February	1,25,170	7.65	15,012	9.73
March	1,38,791	8.49	12,083	7.83
April	1,58,512	9.69	8,019	5.20
May	1,36,347	8.34	9,945	6.45
June	1,25,339	7.66	23,202	15.04
July	1,05,343	6.44	5,088	3.30
August	91,209	5.58	7,506	4.87
September	1,34,575	8.23	5,730	3.72
October	1,53,405	9.38	8,308	5.39
November	1,55,682	9.52	14,078	9.13
December	1,70,904	10.45	21,027	13.63
Yearly trends (1981-84)				
1981	6,11,235	—	1,23,718	—
1982	10,97,325	79.52	1,44,227	16.58
1983	13,67,873	24.65	1,39,075	–3.57
1984	16,35,392	19.56	1,54,227	10.89

TABLE 2.6: Meerut City – Outflow of Commodities Through Railways – 1984 (Spatial Trends)

(Weight : In Tonnes)

Commodities	*States*	*Percentage share of constituent*	*Supplies (in tonnes)*	*Percentage to the total supplies*
Heavy goods				
Jagree	Rajasthan	25.00	10,540	25.42
	West Bengal	75.00		
Khandsari molasses	Orissa	100.00	2,686	6.48
Gawar	Punjab	100.00	60	0.14
Floor	U.P.	100.00	45	0.11
Khandsari	Delhi	100.00	768	1.85
Potato	West Bengal	100.00	126	0.30
Wheat	West Bengal	100.00	1,104	2.66
Miscellaneous goods	Assam	10.00	1,929	4.65
	Bihar	20.01		
	Gujarat	45.00		
	U.P.	24.99		
Fertilizer	J&K	60.00	380	0.92
	Punjab	40.00		
Crushed bone	Maharashtra	100.00	1,732	4.18
Railway material	U.P.	34.97	612	1.48
	Delhi	65.03		
Matting	Haryana	100.00	6	0.01
Waste cotton	West Bengal	100.00	300	0.72
Defence goods	Delhi	85.00	21,181	51.08
	Punjab	6.00		
	U.P.	5.00		
	Maharashtra	3.00		
	Assam	1.00		
Total			41,469	100.00
Light goods				
Printed books	U.P.	50.00	2,926	44.60
	Bihar	19.99		
	M.P.	12.00		
	Rajasthan	8.00		
	Maharashtra	10.01		
Medicines	U.P	74.87	585	8.92
	M.P.	25.13		
Sports goods	Maharashtra	44.97	705	10.75
	U.P.	25.25		
	Tamil Nadu	14.89		
	Haryana	9.93		
	M.P.	4.96		

Contd.

TABLE 2.6 *Contd.*

Commodities	*States*	*Percentage share of constituent*	*Supplies (in tonnes)*	*Percentage to the total supplies*
Handloom cloth	Bihar	25.81	1,170	17.84
	Punjab	21.71		
	U.P.	19.83		
	M.P.	16.41		
	J&K	16.24		
Band goods	Maharashtra	70.09	468	7.13
	Tamil Nadu	14.96		
	Kerala	10.04		
	Rajasthan	4.91		
Defence goods	Punjab	55.67	706	10.76
	Rajasthan	8.92		
	H.P.	4.96		
	Delhi	30.45		
Total			6,560	100.00

Light goods included printed books, medicines, sports goods, handloom cloth, band goods and defence goods. Of the total outward supply of light goods, printed books ranked first and accounted for about 45 per cent of the total outflow under this category. It is followed by the outflow of handloom cloths with about 18 (17.84) per cent of the total light goods. Defence goods and sports goods together accounted for about 22 per cent, medicines nine per cent and band goods seven per cent.

Unlike spatial pattern of heavy goods, light goods were supplied to several States in the country. Printed books were the main items of supply under this category. About 3 thousand tonnes of printed books were supplied to the States of Uttar Pradesh (50%), Bihar (20%), Madhya Pradesh (12%), Maharashtra (10%) and Rajasthan (8%). Supply of these books covered the Hindi-speaking belt of the country. It becomes clear, from these facts, that printing is one of the important functions of the city. Meerut is also an important centre of cotton textile industry. A large quantity of handloom cloth was supplied from the city. About 12 hundred tonnes of handloom cloth was supplied to the States of Bihar (25.81%), Punjab (21.71%), Uttar Pradesh (19.83%), Madhya Pradesh (16.41%) and Jammu and Kashmir (16.24%). From these figures, it is clear that almost whole of the northern India is the consumer market of textile cloth from Meerut city. Higher demand from northern States is perhaps because of the climatic suitability of the handloom cloth in this part of the country. Growing demand of handloom cloth has accelerated the production of cotton textile in the city. Sports goods rank third in terms of supply from Meerut. Supply of sports goods was about 7 hundred tonnes during 1984 and was directed to the States of Maharashtra, Tamil Nadu, Uttar Pradesh, Haryana and Madhya Pradesh. Light defence goods recorded an outflow of about 7 hundred tonnes from Meerut. These goods were supplied to the States of Punjab, Delhi, Rajasthan and Himachal Pradesh. While 600 tonnes of medicines were supplied to the neighbouring States of Uttar Pradesh and Madhya Pradesh, 500 tonnes of band goods were supplied to the distant States like Kerala, Tamil Nadu, Maharashtra and Rajasthan.

Spatial pattern of outflow presents a general tendency of decline towards long distances. However, there are a few exceptions to the above generalisation. Exceptions are due to marked fluctuations in the absolute outflow and quality based demand of selected commodities. There is a positive association between the long distances and light weight goods. In case of heavy goods, matting, jaggery and defence goods were supplied to short distances whereas khandsari, fertilizer, crushed bone and miscellaneous goods to medium distances. On the other hand, light goods, like, medicines, printed books and light defence goods, were supplied to the short distances, sports goods and handloom fabrics to the medium distances and sports goods and band goods to the long distances (Fig. 2.6).

Temporal outflow (Table 2.7) of commodities reveals a declining trend in case of heavy goods and increasing trend in case of light goods. Over a period of a decade outflow through railways registered a decline of 59 thousand tonnes, i.e., from about one lakh tonnes in 1975 to about 41,000 tonnes in 1984. The outflow of light goods was about 6 (5.8) thousand tonnes at the beginning (1975) which rose to the level of about 7 thousand tonnes by the turn of 1984. There are marked variations in the flow patterns of the two types of goods (light as well as heavy goods). As a result, the overall outflow (heavy + light goods) registered a decline of 56 thousand tonnes during 1975-84. Rate of decline ranged between five (4.68 per cent in 1979-80) to 24 (23.62 per cent in 1978-79) per cent during the period under study. The monthly trend of the supply of heavy goods reveals that low amount of outflow between July to October, moderate outflow between November to February and high outflow between March to June. Movement of heavy goods is adversely influenced by the rainy season (Fig. 2.7). On the contrary, light goods registered a period of low outflow during April to June, a period of moderate outflow during January to March and high outflow during July to December. Thus, the trends of monthly flow for the two types of goods are not identical as they present marked variations. Analysis further explains that short and medium range supply of heavy goods through railways faces strong competition with the highways which have improved supply conditions significantly (Fig. 2.7).

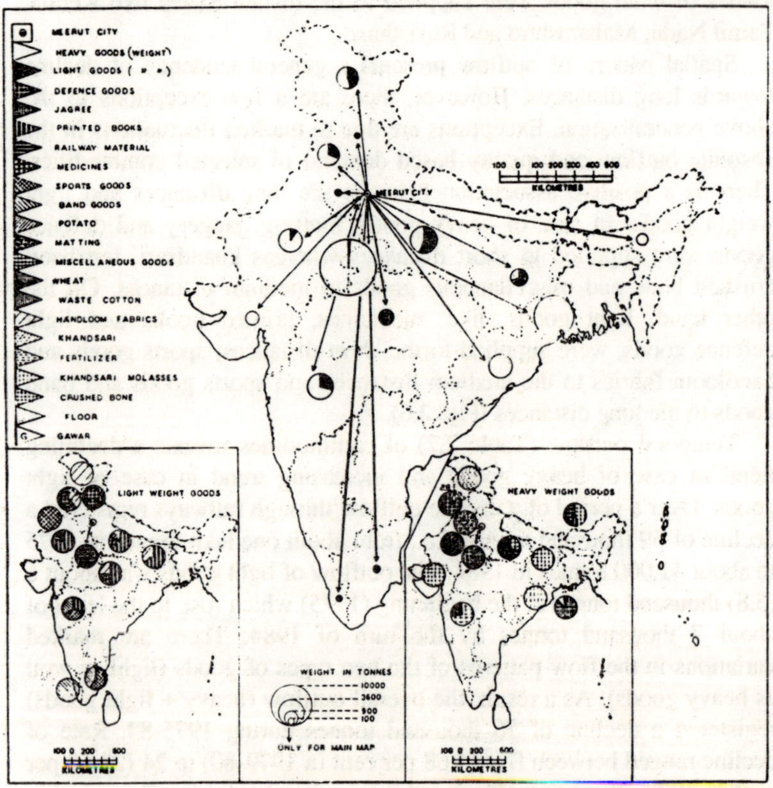

FIG. 2.6: Meerut city — Outflow of commodities through railways — 1984.

TABLE 2.7: Meerut City — Temporal Trends of Outflow Through Railways

(Weight : In Tonnes)

Monthly trends (1984)				*Annual trends (1975-84)*			
Months	*Light goods*	*Heavy goods*	*Share of monthly supply*	*Years*	*Light goods*	*Heavy goods*	*Growth rate*
January	534	3,211	7.80	1975	5,844	1,00,390	—
February	420	2,724	6.55	1976	6,841	90,921	–7.97
March	533	8,203	18.19	1977	6,249	1,04,711	13.50
April	376	3,800	8.69	1978	6,516	84,531	–17.95
May	308	5,495	12.08	1979	6,737	62,805	–23.62
June	362	5,282	11.75	1980	6,240	60,046	–4.68
July	686	1,761	5.09	1981	6,181	60,997	1.34
August	498	1,421	4.00	1982	6,466	52,834	–11.72
September	568	841	2.93	1983	5,930	46,749	–11.16
October	954	975	4.02	1984	6,560	41,469	–8.83
November	731	3,994	9.84				
December	590	3,762	9.06				

It is, thus, possible to generalise that weight of outflow tends to decline with the distance offering scope for light goods to move longer distances and heavy goods to limit their supply to shorter distances, i.e., to the northern and central parts of the country. Further, weight of a commodity and its quality also affects the range of outflow. For instance, band goods and sports goods are considered qualitatively superior productions as such they are in demand from distant areas like Kerala and Tamil Nadu.

Trends of Earning through Railways

Movement of men and material acts as a source of earning to the transport sector which plays a vital role in the development of economy and the region. Behaviour of booking (goods) has a direct bearing on earnings. Variation in the monthly booking explains effect of seasonality whereas variation in annual booking explains growth trends of earnings. Since details of booking through highways were not available, trends of earning through goods booking could be analysed only in case of railways. Earnings have been analysed (Table 2.8) for a period of 10 years (1975-84). Annual trend of earning reveals sharp rise in light goods and moderate to low rise in heavy goods. During 1975 light goods recorded the earnings of about Rs. 10 (10.2) lakhs, which rose to the level of about Rs. 22 (21.5) lakh by 1984. While earnings have risen over 100 per cent, weight rose only by 60 per cent. Similar trend was recorded in case of heavy goods, which earned Rs. 21 lakhs in 1975 and Rs. 47 lakhs in 1984. Thus, earnings of heavy goods rose more than 100 per cent within a decade whereas the weight registered a fall by about 48 (47.57) per cent.

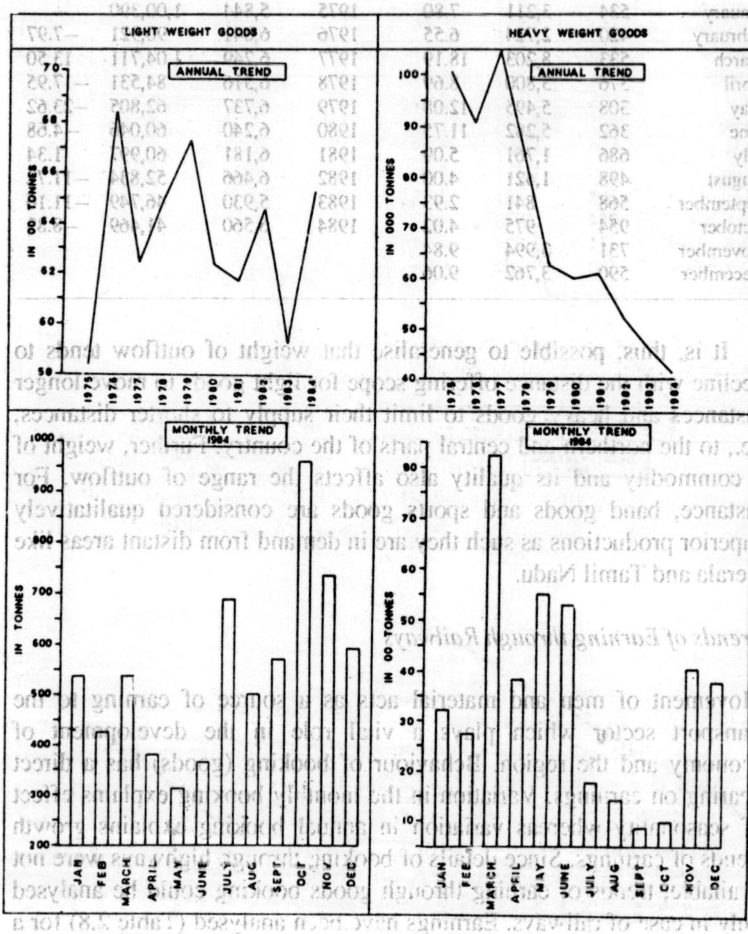

FIG. 2.7: Meerut city — Outflow trends of commodities through railways.

Unlike annual earnings, monthly earnings (Fig. 2.8) present marked variations. In case of light goods, earnings ranged between Rs. 130 thousand (Rs. 1.30 lakhs) in May to Rs. 230 thousand (Rs.2.30 lakhs) in December. While January to June experienced low earnings, July to December recorded high earnings. On the contrary, earnings through heavy goods ranged between Rs. 25 lakhs (in September) and Rs. 65 lakhs (in May). The period between June to September (rainy season) experienced relatively low earnings whereas period between October to May experienced high earnings. There are, as such, large variations in the trends of monthly earnings.

TABLE 2.8: Meerut City — Trends of Railway Earnings Through Goods Booking

(In Million Rupees)

Monthly trends (1984)					*Annual trends (1975-84)*				
Months	*Light weight goods*	*Heavy weight goods*	*Total*	*Percent-age to total*	*Year*	*Light weight goods*	*Heavy weight goods*	*Total*	*Growth rate*
January	0.15	4.74	4.89	8.98	1975	1.03	20.46	21.49	—
February	0.14	4.10	4.24	7.79	1976	1.16	32.80	33.96	58.03
March	0.16	5.88	6.04	11.09	1977	1.24	29.37	30.61	–9.86
April	0.16	5.44	5.60	10.28	1978	1.45	28.31	29.76	–2.77
May	0.13	6.34	6.47	11.89	1979	1.15	27.73	28.88	–2.96
June	0.16	3.33	3.49	6.41	1980	1.33	36.06	37.39	29.47
July	0.19	2.61	2.80	5.15	1981	1.43	43.20	44.63	19.36
August	0.19	4.36	4.55	8.36	1982	1.90	57.40	59.30	32.87
September	0.17	2.45	2.62	4.81	1983	2.05	34.84	36.89	–37.79
October	0.22	3.85	4.07	7.47	1984	2.10	52.34	54.44	47.57
November	0.24	4.33	4.57	8.39					
December	0.19	4.91	5.10	9.37					

Results suggest that despite increase in railway earnings, increase in the weight of booking could not be observed. While goods booking increased for short and medium ranges, goods booking for long range remained quite low. This relates to the linkages over long distances. Further rise in railway booking is possible only through the improved system of booking and delivery.

The above facts explain that there is a positive association between the weight of railway booking and its earnings. However, the two trends are not co-linear throughout the period under review. Marked variations are partly due to the factor of distance and partly due to the type of booking available, i.e., (i) charges paid at the time of booking and (ii) charges to be paid at the time of goods delivery.

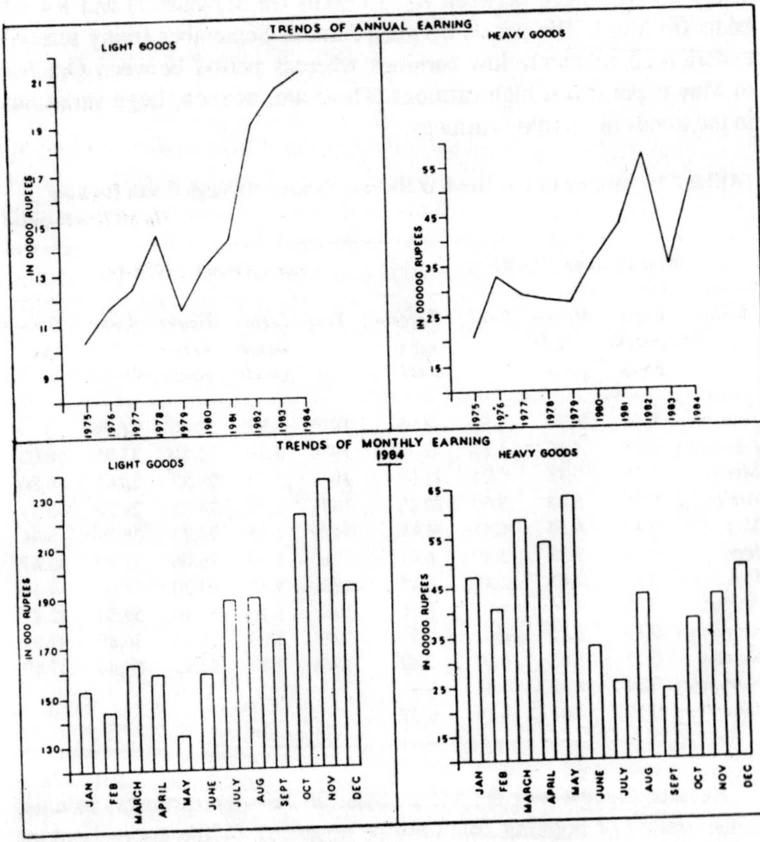

FIG. 2.8: Meerut city – Trends of earning through railway bookings.

Nature of Surplus and Deficit Commodities

Having analysed flow relations of Meerut, it was essential to work out the nature of surplus and deficit commodities and inter-dependence of the constituent areal units. While term surplus refers to the excess that remains after the requirements, term deficit refers to the quantity by which a production runs short of the requirements. In the light of the coverage of these terms, following procedure has been adopted to obtain the desired results under the given constraints of data availability:

(i) to work out inflow-outflow differentials, where
(ii) the positive differentials denote surplus and negative denote deficit.

This exercise relates (Table 2.9) to the analysis of 28 commodities for the years 1981, 1982 and 1983 and 38 commodities for the year 1984. Since the money value of the commodities was not available, surplus and deficits have been calculated in terms of weight only. Results reveal that out of the total commodities, 13 recorded surplus in 1981, 15 in 1982, 14 in 1983 and 25 in 1984. Contrary to this, deficits were recorded in as many as 13 commodities in 1981, 12 in 1982, 13 in 1983 and 13 in 1984. Temporal patterns of commodities reveal three main trends of commodities, viz.:

(i) commodities registering surplus in all the years (1981 to 1984).
(ii) commodities registering deficit in all the years, and
(iii) commodities registering mixed pattern of deficit and surplus.

The first trend was observed in case of 14 commodities, the second in case of 10 commodities and the third in case of 4 commodities.

Surplus Commodities

Among the surplus commodities food items, sweets, textile cloth, milk, vegetables, milk-made items, livestock feed, fruits, crushed bone, sports goods, band goods, waste cotton, cutlery and matting reported surplus in all the years from 1981 to 1984. Whereas, printed books recorded surplus during 1982 and 1983 and railway material during 1981 and 1984. During 1984, 25 commodities were reported as surplus. The quantity of surplus ranged from six tonnes (in case of matting) to 3.04 lakh tonnes (in case of food items). Meerut, being one of the developed districts of the country, has surplus in agricultural, horticultural, livestock feed and dairy and other industrial products. In 1984 registered a weight of about 6.13 lakh tonnes as surplus. Of the

TABLE 2.9: Nature of Surplus and Deficit Commodities of Meerut City — 1984

(Weight : In Tonnes)

Commodities	Surplus	Percentage	Deficit	Percentage
1. Printed books	—	—	146.00	0.07
2. Medicines	—	—	583.00	0.03
3. Sports goods	1,577.63	0.26	—	—
4. Waste cotton	300.00	0.05	—	—
5. Textiles	49,899.17	8.19	—	—
6. Handloom machine	—	—	384.00	0.02
7. Band goods	468.00	0.08	—	—
8. Petroleum	—	—	87,229.00	4.50
9. Coal	—	—	97,700.00	5.04
10. Firewood and fuel	—	—	5,65,142.00	29.16
11. Timber	—	—	26,176.50	1.35
12. Construction goods	—	—	7,07,190.28	36.48
13. Fertilizers	—	—	2,12,188.00	10.95
14. Sweets	37,126.70	6.10	—	—
15. Food items	3,04,533.42	50.01	—	—
16. Vegetables	15,437.10	2.53	—	—
17. Fruits	5,441.20	0.89	—	—
18. Livestock feed	6,159.50	1.01	—	—
19. Crushed bone	1,732.00	0.28	—	—
20. Leather	—	—	715.98	0.04
21. Matting	6.00	0.001	—	—
22. Milk	20,829.42	3.42	—	—
23. Milk-made items	11,668.35	1.92	—	—
24. Cutlery	87.03	0.01	—	—
25. Metal	—	—	83,605.68	4.31
26. Railway material	152.00	0.02	—	—
27. Defence goods	—	—	1,02,251.00	5.27
28. Sugar	57,000.00	9.36	—	—
29. Rubber goods	72,900.00	11.97	—	—
30. Transformer	1,500.00	0.25	—	—
31. Surgical goods	600.00	0.10	—	—
32. Agricultural implements	9,125.00	1.50	—	—
33. Solder-wire	420.00	0.07	—	—
34. *Gobar* gas plant implements	1,825.00	0.30	—	—
35. Liquor	5,400.00	0.89	—	—
36. Pharmaceutical sugar cube	1,800.00	0.29	—	—
37. Chemicals	3,000.00	0.49	—	—
38. Other commodities	—	—	54,987.21	2.84
Total	6,08,987.52	100.00	19,38,298.65	100.00

Note: Further details about surplus and deficit of 1981, 1982 and 1983 see Appendix-1.

total surplus, share of agricultural surplus was nearly 61 (60.54) per cent. Among the surplus commodities, food items ranked first accounting for 50 per cent of the (3 lakh tonnes) total. Milk and vegetables were next in order of importance with respective surpluses as 21 (20.8) and 15 (15.4) thousand tonnes. Similarly livestock feed and fruits had a surplus of 6 (6.1) and 5 (5.4) thousand tonnes, respectively.

Among surplus industrial products, rubber goods, sugar, textile cloth, milk-made items and agricultural implements are the most important. Rubber goods recorded a surplus of 73 (72.90) thousand tonnes, sugar 57 thousand tonnes, textiles about 50 thousand tonnes, milk 21 (20.82) thousand tonnes, milk-made items 12 (11.66) thousand tonnes and agricultural implements 9 (9.12) thousand tonnes. The share of these commodities in overall surplus was 12 (11.97), nine (9.36), eight (8.19), three (3.42), two (1.92) and 1.50 per cent, respectively. Crushed bone and sports goods are next in order of industrial surplus, the respective weights being 1.73 and 1.57 thousand tonnes. Commodities with low quantity of surplus (below 500 tonnes) during 1984 were band goods, waste cotton, solder-wire, railway material, cutlery and matting during 1984.

Deficit Commodities

There were 13 commodities showing a net deficit of 1.94 million tonnes during 1984. The quantity of deficit ranged from 146.00 tonnes (printed books) to 7.07 lakh tonnes (construction goods). Among the deficit commodities, construction goods and firewood and fuel accounted for more than 65 per cent of the total deficit and as such were grouped under the commodities with very high deficit. High amount of deficits was reported in case of fertilizers (2.12 lakh tonnes) and defence goods (1.02 lakh tonnes); the share in the total deficit being about 16 per cent. Coal, petroleum and metal were classed as commodities of low deficit, quantity being 98 (97.7), 87 (87.2), and 84 (83.6) thousand tonnes respectively. Together these commodities accounted for about 14 per cent of the total deficit. Finally, commodities, like, timber, leather, medicines, handloom machines, printed books and other commodities recorded the very low level of deficit. A net deficit of about 83,000 tonnes accounting for about four (4.28) per cent was recorded in case of these commodities during 1984.

These facts reveal that:

(i) Overall surplus is less than the deficit. It is mainly due to the receipt of heavy items, like, cement, coal, fertilizer and firewood. It would have been better to compare the two in terms of value rather than weight. However, due to the absence of necessary data, other alternatives were not available,

(ii) Agricultural products dominate the surplus. Nonetheless industrial products and other items are also significant in regional context. It is, however, observed that agriculture has been largely mechanised and modernised, as such it recorded higher surplus. On the contrary, industry continues to remain traditional to semi-modern in most of the productions. As a result limited quantity of industrial surplus is generated,

(iii) Deficit is largely associated with those products which are not produced in the region. Meerut, therefore, depends upon States of Bihar, Madhya Pradesh, Gujarat, Assam and Uttar Pradesh for commodities like coal, fertilizer, petroleum and wood, and

(iv) It is, however, realised that efforts are needed to generate more agricultural as well as industrial surplus to bridge the prevailing functional and income gaps.

Functional Gaps and Income Inequalities

Study of functional gaps and prevailing income inequality offers a base for developing plans to combat the imbalances within a region. While functional gap refers to the existing variation in the distribution, income inequality refers to the variation in its earnings. Functional gaps and income inequalities have been analysed in cases of surplus commodities alone, because region pays for its deficit whereas it earns from its surplus. There were 25 surplus commodities in Meerut during 1984. Of these, 16 have been included in this exercise. Selection of commodities is based on the assumption that block, as an areal units, contributes the average surplus and income to the process of development in the district. A separate account for functional gaps and income inequalities is given below:

Functional Gaps

Most of the studies on functional gaps have been largely concerned with the identification of distributional gaps in socio-economic facilities and amenities. Present study, however, deviates from the theme as it concerns the identification of gaps in the functional establishments (units) of selected surplus commodities. The purpose of the exercise is to analyse the distribution (availability as well as gaps) of surplus generating factor in the district. While ubiquitous distribution is an ideal condition for growth, clustered distribution is the major constraint for balanced growth of the regions. Methods used for the calculation of functional availability and gaps are given as under:

$$\text{Availability} = \frac{\text{Number of blocks with a function X}}{\text{Total number of blocks}} \times 100$$

and,

$$\text{Gaps} = \frac{\text{Number of blocks without a function X}}{\text{Total number of blocks}} \times 100$$

Functional availability is marked with the presence of a registered unit of selected industries whereas functional gap is marked with the absence of that unit.

Available Units: Surplus commodities noted above have in all 1,011 registered industrial units. Number of industrial units ranged between a single unit (milk-made items) to 241 (sports goods) units. Large number of industrial units (more than 100) were recorded in case of sports goods, metal works and printing works. Medium number of industrial units were recorded in cases of leather goods, sweets and cutlery. While low number of industrial units (25-50) was recorded in construction goods, textiles and timber; very low number (below 25) of industrial units was recorded in medicines, band goods, food items, milk and milk-made items.

Spatial distribution of industries reveals that over 80 per cent of the total industrial units were concentrated in Meerut city alone. On the contrary, blocks of Rajpura, Jani, Saroorpur, Binoli and Chhaprauli had no registered industrial unit. Blocks of Parikshhatgarh and Khekra had one, Kharkhauda two and Rohta, Hastinapur, Machhra and Daurala three types of industrial units. While Mawana recorded four types of industrial combination, Baghpat recorded five types. Relatively more combinations of industry were recorded in cases of Baraut (six) and Sardhana (eight); number of registered units being 28 and 86, respectively.

Spread of industries reveals that out of 16, six commodities were located at one place, two commodities at two places and three commodities at three places. Thus, 11 commodities have shown very limited spread. Contrary to this, surplus commodities like leather goods, construction goods, timber, metal and sweets (sugar) were distributed over five to 10 blocks in the district. It is, thus, evident that most of the industries have a tendency to concentrate (Table 2.10).

Employment in the Units: The importance of a function at a given place and time is judged by the level of employment it generates. In the absence of details about real employment, employment in the registered industrial units has been analysed. These industries accounted for an employment of 25,879 persons, the range of employment being 13 (milk) to 10,397 (sweets) persons. Relatively higher employment (over 2,500 persons) were observed in metal, textiles and other commodities whereas medium level of employment (600 – 1,800 persons) was recorded in cases of sports goods, construction goods and printing works. While low level of employment was recorded in leather goods, cutlery, timber works, milk-made items, band goods and medicines, very low level (below 100 persons) of employment was recorded in food items, livestock feed, and milk (Table 2.11).

Gaps in the Distribution (Figs. 2.9, 2.10, 2.11 and 2.12): Process of concentration has left large gaps in industrial organisations of the space which has adversely affected the process of development. Extent of functional gap ranged between 44 to 94 per cent in district Meerut. An extremely large (94.44%) gap existed in functions like processing of milk and milk-made items, manufacturing of livestock feed, sports goods, band goods and cutlery. These functions were localised only in the Meerut city. Other commodities and processing of food items also recorded fairly large (88.88%) gap as they were concentrated at Sardhana and Baraut apart from Meerut. Large gaps (83.33%) existed in the manufacturing of medicines, textiles and printing works. These establishments have developed in places, like, Meerut, Sardhana, Baraut and Hastinapur. Manufacturing of leather goods, construction goods and timber presented medium gap (between 61-72%) as these units were recorded in the blocks of Rohta, Machhra, Sardhana, Baghpat, Mawana, Daurala, Baraut and Kharkhauda in addition to Meerut. On the contrary, low level of functional gaps (44.55%) was recorded in metal works and sweets. Establishments related to these functions were recorded in as many as eight to ten blocks in the district (Table 2.10).

TABLE 2.10: Meerut District — Spatial Patterns of Registered Functional Units — 1984 (Gaps and Availability)

(In Numbers)

Blocks	Commodities									
	Sweets	*Medicines*	*Food items*	*Textiles*	*Timber*	*Construction goods*	*Leather goods*	*Metal*	*Printing*	*Other commodities*
1. Meerut	24	11	2	26	24	27	30	204	98	96
2. Jani	—	—	—	—	—	—	—	—	—	—
3. Rohta	2	—	—	—	—	1	2	—	—	—
4. Rajpura	—	—	—	—	—	—	—	—	—	—
5. Kharkhauda	3	—	—	—	2	—	—	—	—	—
6. Mawana	7	—	—	—	1	4	—	3	—	—
7. Hastinapur	1	—	—	1	—	—	—	1	—	—
8. Parikshhatgarh	5	—	—	—	—	—	—	—	—	—
9. Machhra	4	—	—	—	—	1	1	-	—	—
10. Sardhana	8	—	1	10	3	6	53	1	1	—
11. Saroorpur	—	—	—	—	—	—	—	-	—	—
12. Binoli	—	—	—	—	—	—	—	—	—	—
13. Daurala	2	—	—	—	1	—	—	1	—	—
14. Baghpat	3	1	—	—	1	—	1	3	—	—
15. Baraut	—	3	—	—	1	1	—	18	4	1
16. Chhaprauli	—	—	—	—	—	—	—	—	—	—
17. Khekra	—	—	—	—	—	—	—	1	—	—
Total	59	15	3	37	33	40	87	235	103	97

Note: Blank space in the columns express functional gaps. Registered industries of milk, milk-made items, livestock feed, sports goods, band goods and cutlery are located only in the Meerut block. The number of functional units for these industries was 2, 1, 3, 241, 5 and 50, respectively. Since these industries have no establishment in other parts of the district, they could not be displayed in the Table.

TABLE 2.11: Meerut District — Employment and Income in Registered Units — 1984

Commodities	*Unit*	*Employment*		*Annual production capacity (in lakh rupees)*		*Per unit income (in lakh rupees)*	*Per worker income (in lakh rupees)*
		In Number	*Percentage to total*	*Rupees*	*Percentage to total*		
1. Milk	2	13	0.05	4.10	0.03	2.05	0.31
2. Milk-made items	1	184	0.71	750.00	6.03	750.00	4.08
3. Sweets	59	10,397	40.18	2,660.60	21.38	45.09	0.25
4.Medicines	15	138	0.53	42.68	0.34	2.84	0.31
5. Livestock feed	3	25	0.10	26.14	0.21	8.71	1.04
6. Food items	3	83	0.32	361.19	2.90	120.40	4.35
7. Textiles	37	2,585	9.99	1,545.69	12.42	41.77	0.60
8. Timber	33	194	0.75	43.80	0.35	1.33	0.22
9. Sports goods	241	1,746	6.75	340.64	2.74	1.41	0.19
10. Band goods	5	141	0.54	22.17	0.18	4.43	0.16
11. Construction goods	40	1,041	4.02	75.27	0.60	1.88	0.07
12. Leather goods	87	300	1.16	75.06	0.60	0.86	0.25
13. Metal	235	2,886	11.15	988.72	7.95	4.21	0.34
14. Cutlery	50	279	1.08	44.81	0.36	0.90	0.16
15. Printing	103	604	2.33	267.90	2.15	2.60	0.44
16. Other commodities	97	5,263	20.34	5,197.29	41.76	53.58	0.98
Total	1,011	25,879	100.00	12,446.06	100.00	1,042.06	

Source: Computed from *Distt. Industries Bulletin*, Meerut — 1985.

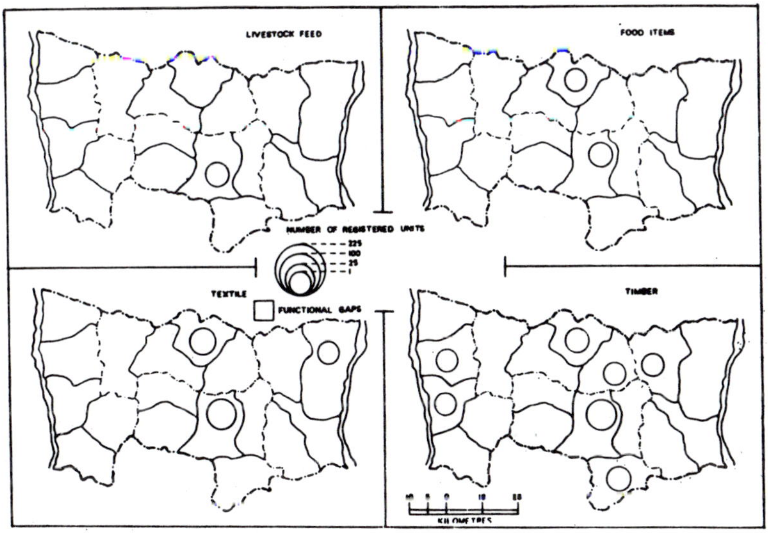

FIG. 2.9: Meerut — Functional availability and gaps — livestock feed, food items, textile and timber.

Income Inequality

Inequality is an important measure of variation and explains the extent of income gap in the development. Due to non-availability of data on income, informations about the production capacity (in money terms) of 16 commodities[4] were collected and used in this exercise. It is based on the supposition that production capacity in money terms indicates relative levels of income generation by each industrial unit. As such production capacity explains levels of income for industrial products under study. Details of these informations were taken from the Directory of Industries, Meerut, 1984. Income inequality, in this case, refers to the observed variations in the per unit production capacity (in money terms) of each industry as well as per unit production capacity (in money terms) of industrial worker in case of 16 industrial products. For the sake of convenience, variations have been worked out for all units together and for large and small units separately. This section deals with the distributional aspects of income (production capacity in money terms) across unit industries and unit areas. A separate discussion about income distribution and income inequality is given below:

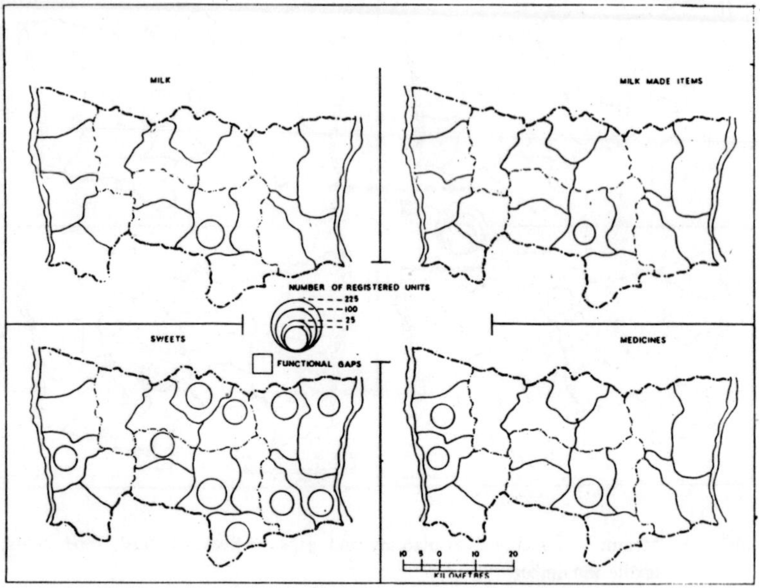

FIG. 2.10: Meerut — Functional availability and gaps — milk, milk-made items, sweets and medicines.

Distribution of Industrial Income: Industrial income presents significant variations among groups of industrial products and their areal distribution. Income (i.e., Rs. 1,244.60 million) generated by 16 industrial products ranged between Rs. 4.10 lakhs (milk) to Rs. 266 million (sweets). Share of different industrial products presents marked variation in the industrial income of the district. It is about three (2.67) per cent by dairy works, band goods, livestock feed, medicines, timber, cutlery, construction goods and leather goods. 34 (34.19) per cent by sport goods, food items, printing works, manufacturing of milk-made items, metal works and textiles. Remaining 63 (63.14) per cent was contributed by sweets and other commodities. It is, therefore, clear that 11 out of 16 products (according to rank) contributed as much as nearly 90 (89.54) per cent of the total industrial income of the district (Table 2.12).

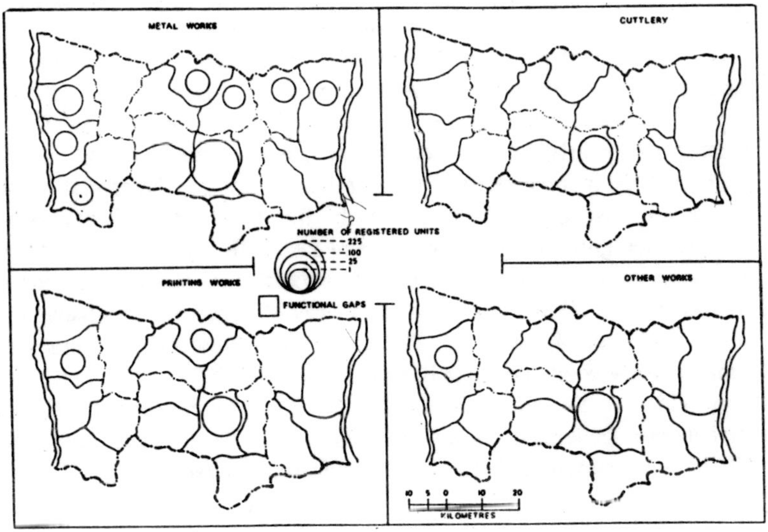

FIG. 2.11: Meerut — Functional availability and gaps — metal works, cutlery, printing works and other works.

Disparities have been observed in the income generation by various areal units. Nearly 80 (79.69) per cent of the total industrial income generates from Meerut city alone. Next in order of areal importance is Mawana which accounts for about 10 (9.65) per cent of the industrial income. Hastinapur ranks third and contributes nearly four (4.05) per cent of the income. Blocks of Daurala and Sardhana are largely at the same level of income generation with their respective shares as 2.5 (2.47) and two (2.07) per cent. Remaining parts of the district share about two per cent of the total income.

There are marked distributional gaps in terms of dispersal of industrial units and income generation at the block level. For instance, income through medicines amounted to Rs. 43 (42.68) lakhs whereas through sweets Rs. 266 million. Areal responses to the product-wise industrial income through medicines are: Baraut accounted for about three per cent, Baghpat about two per cent and Meerut about 95 per cent. Similarly, timber manufacturing contributed about Rs. 44 (Rs. 43.80) lakhs to the industrial income. The shares of various units were: over 75 per cent Meerut, nine (9.13) per cent Daurala, seven (7.40) per cent Sardhana, four (3.88) per cent Baraut, two (2.28) per cent Mawana and remaining below two per cent from Kharkhauda. Unlike income through medicines and timber, manufacturing of leather goods contributed a sum of Rs. 75 (Rs. 75.06) lakhs to the industrial income. Of the income generated through leather goods, Sardhana accounted for a maximum of over 53 (53.29) per cent followed by Meerut, i.e., nearly 45 per cent. Baghpat, Rohta and Machhra shared the rest of the income (1.73%) through leather goods. The spatial trend of construction goods follows closely the trend of leather goods. Construction goods contributed a sum of Rs. 75 (Rs. 75.26) lakhs to the industrial income of the district. The proportionate share of different areal units was 55 per cent Meerut, 17 (17.40) per cent Sardhana, 13 (13.29) per cent Mawana, 13 (13.28) per cent Rohta and Machhra and remaining about one per cent Baraut. About Rs. 27 (Rs. 26.79) million were added to the industrial income through printing works. Of the total income generated through printing, nearly 98 per cent was from Meerut and remaining about two per cent from Baraut and Sardhana.

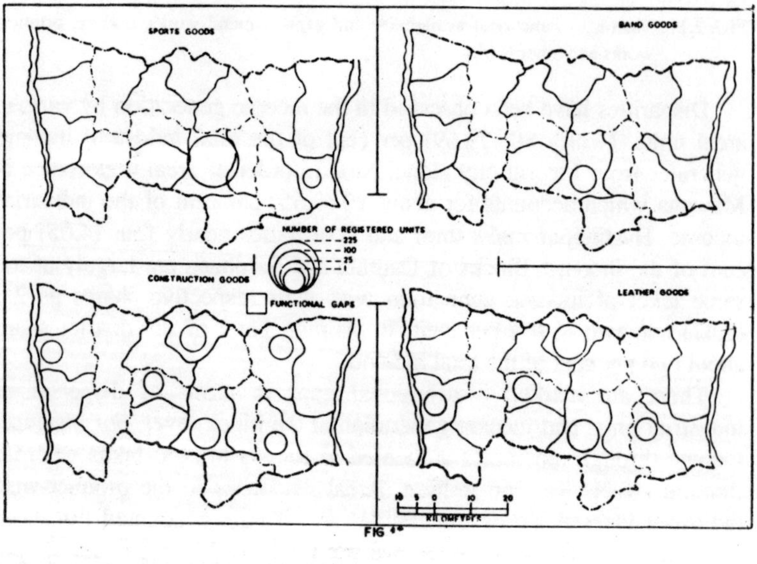

FIG. 2.12: Meerut — Functional availability and gaps — sports goods, band goods, construction goods and leather goods.

TABLE 2.12: Meerut District — Spatial Patterns of Income from Registered Industrial Units — 1984

(Income in '00000 Rs.)

Blocks	*Commodities*									
	Sweets	*Medicines*	*Food items*	*Textiles*	*Timber*	*Construction goods*	*Leather goods*	*Metal*	*Printing*	*Other commodities*
1. Meerut	844.23	40.59	359.19	1,000.44	32.99	41.55	33.76	919.57	261.31	5,196.66
2. Jani	—	—	—	—	—	—	—	—	—	—
3. Rohta	6.40	—	—	—	—	5.00	0.50	—	—	—
4. Rajpura	—	—	—	—	—	—	—	—	—	—
5. Kharkhauda	5.35	—	—	—	0.72	—	—	—	—	—
6. Mawana	1,187.87	—	—	—	1.00	10.00	—	2.82	—	—
7. Hastinapur	1.10	—	—	500.00	—	—	—	3.00	—	—
8. Parikshhatgarh	5.35	—	—	—	—	—	—	—	—	—
9. Machhra	7.50	—	—	—	—	5.00	0.20	—	—	—
10. Sardhana	147.80	—	2.00	45.25	3.24	13.10	40.00	4.64	2	—
11. Saroorpur	—	—	—	—	—	—	—	—	—	—
12. Binoli	—	—	—	—	—	—	—	—	—	—
13. Daurala	302.50	—	—	—	4.00	—	—	0.50	—	—
14. Baghpat	152.50	0.89	—	—	0.15	—	0.60	6.30	—	—
15. Baraut	—	1.20	—	—	1.70	0.62	—	51.14	4.59	0.63
16. Chhaprauli	—	—	—	—	—	—	—	—	—	—
17. Khekra	—	—	—	—	—	—	—	0.75	—	—
Total	2,660.60	42.68	361.19	1,545.69	43.80	75.27	75.06	988.72	267.90	5,197.29

Note: Incomes through milk, milk-made items, livestock feed, sports goods, band goods and cutlery have been recorded only in case of Meerut, their incomes being Rs. 4.10, Rs. 750.00, Rs. 26.14, Rs. 340.64, Rs. 22.17 and Rs. 44.81 lakhs, respectively.

Metal works (Rs. 98.9 million) contributed more than three times to that of income through printing works. Of the total income generated through printing works, 93 per cent was from Meerut whereas remaining seven per cent was shared by the blocks of Baraut, Baghpat, Sardhana, Hastinapur, Mawana and Khekra. Income through cotton textiles was still higher, i.e., about Rs. 155 million (Rs. 154.56 million). Of the total income through textiles, Meerut contributed nearly 65 per cent (64.72%), Hastinapur 32 (32.35) per cent and Sardhana about three (2.93) per cent. Sweets shared the maximum income, i.e., Rs. 266 million. The areal account of the sweets was nearly 45 (44.65) per cent for Mawana, 32 (31.74) per cent for Meerut, 11 (11.37) per cent Daurala and about six (5.55) per cent each for Baghpat and Sardhana. Remaining less than one per cent was shared by Rohta, Kharkhauda, Parikshhatgarh, Machhra and Hastinapur. It is, thus, established that industrial income displays large inequalities both in sectoral and spatial terms. There is a need to integrate the function of income generation across the commodities and areal units.

Variation in Industrial Income: The foregoing facts explain that there are marked areal and industrial gaps in the income (Table 2.13). Per unit area and industrial income is discussed in terms variation. Per unit capacity in money terms ranged from Rs. 86 thousand in leather goods to Rs. 75 million in milk-made items. Out of 16 commodities, 11 registered their capacity between Rs. 1 to 8 lakhs. Remaining five units reported income between Rs. 4 to 75 million. The income per worker, however, ranged from Rs. 7,000 in construction goods to about Rs. 4 (Rs. 4.08) lakhs in milk-made items. Per worker income has been lower in almost all functional units except in the processing of food items, livestock feed and milk-made items. The income (production capacity) of all units was about Rs. 1245 (Rs. 1244.60) million. Of the production capacity, large units accounted for more than 84 (84.48) per cent and that of small units about 16 (15.51) per cent. Total number of registered functional units was 1,011. Of all the units, number of large units accounted for about 19 (19.48) per cent and that of small units 81 (80.51) per cent. While average income (capacity) per unit was Rs. 1.23 million for 16 commodities together; average income for the two groups was Rs. 6.51 million. It is, thus, clear that most of the commodities (12) have small units with low income (production capacity). On the contrary, only four commodity groups with large functional units have high income. Average for all units is, therefore, much lower compared to the average for the specific groups. Marked variations in the income (production capacity) have led to the higher standard deviation. For instance, standard deviation for all units was 194, for large units 1,897.14 and for small units 37.98. Deviation for specific units was 179 for all commodities, 275 for the commodities of large units and 2.183 for commodities of small units. Income (production capacity), thus, varies with the size of the units. While inequality remains low with small units, it rises up sharply with the addition of medium or large units. Income (production capacity) of large units was, therefore, significantly higher compared to the income (capacity) of small units. There are two distinct income trends, viz., sharp variations in the per unit income (production capacity) and relatively lesser variation in the income per worker (production capacity). While there are little variations in the income per worker (capacity of per worker), there are sharp variations in the size and the income per unit. Income inequality is, therefore, due to the concentration of large industrial units at a few locations. Besides concentration, the absence of functions over large areas have led to the income inequality in the process of development. Large scale sprout of household industrial units may possibly bridge the income inequality and remove prevailing functional gaps. Lack of efficient co-ordination, between the supply of raw material and production and to that of distribution (market) mechanisms, acts as major constraint in the process of industrial development. Linkages need to be improved so as to observe the quick supply and arrival of surplus and deficit commodities.

TABLE 2.13: Income Inequality in Registered Functional Units

Inequality group	*No. of Units*	*Total income (in lakhs Rs.)*	*Mean Income*	*Standard Deviation*
All units	1,011	12,446.06	12.31	194.00
Major units*	197	10,514.77	53.37	1,897.14
Small units**	814	1,931.29	2.37	37.98
Per unit — All unit types	16	1,042.06	65.12	179.56
Per unit — Major unit types	5	1,010.84	202.16	275.42
Per unit — Small unit types	11	31.22	2.83	2.1834

* Major units include milk-made items, sweets, food items, textiles and other commodities.

** Small units include milk, medicines, livestock feed, timber, sports goods, band goods, construction goods, leather goods, metal, cutlery and printing.

The following generalisation emerge from the foregoing discussion :

(i) There are marked variations in the income levels of all units together,

(ii) In case of large units income variations are very sharp compared to the small units,

(iii) Since the share of large industrial units is greater than the share of small units in the overall income, inequalities tend to remain high in all commodity groups.

Conclusion

The above facts may be summarised below :

(i) There is an annual turnover of about 3 (3.16) million tonnes of commodities from Meerut. Of these about 2½ (2.52) million tonnes were recorded as inflow and remaining about 6½ tonnes (6.44 lakh tonnes) as outflow. About 5½ (5.44) lakh tonnes (i.e., about 22%) of arrivals are used for city consumption. The share of inflow through railways was nearly 28 (28.11) per cent whereas that of outflow over seven (7.45) per cent. Similarly inflow through highways accounted for about 72 (71.89) per cent of the total arrivals and over 92 (92.55) per cent of the total supplies from Meerut city,

(ii) While agricultural products (57.52%) dominate the outflow, construction goods, fertilizer, coal, fuel and petroleum dominate (66.33%) the inflow. Deficit relates to those commodities which are not produced locally. Coal, fertilizer, petroleum and wood are the chief deficit commodities in the city. Quantity of deficit is more than that of the surplus, the difference being that of about 1.32 million tonnes. This is possibly because of the arrival of the weight gaining commodities in the city and inadequate data about outflow of commodities. Results would have been sharp if flows were analysed in terms of money instead of weight. Paucity of required data left with no alternative except to analyse results in terms of weight, and

(iii) Since most of the functions tend to concentrate in a selected location, there are marked areal gaps in the functional availability. Extent of functional gaps ranges between 66 to 85 per cent. While functions of agricultural products are ubiquitously distributed in the district, functions of industrial products are concentrated only in a few locations. As many as six out of 16 commodities are located in Meerut alone, five commodities at two places outside Meerut and remaining five commodities recorded relatively fair distribution. It is accepted that all the functions cannot be made available to all the locations. However, it is desired that medium level functions be made accessible to all points within a reasonable range and at a minimum cost level. Functional concentration has correspondingly caused income inequalities in all the functions. Nearly 80 per cent of the industrial income generates from Meerut city. Besides Meerut, Mawana, Hastinapur, Sardhana and Baraut are important in terms of income generation in industrial sector. Income inequalities are sharp in cases of big units compared to small units. The present set-up is likely to contribute more towards increasing inequalities with the pace of time.

NOTES

1. While light goods are sent through railway parcel, heavy goods are sent through railway bookings. The commodities arriving through railway parcel include printed books, medicines, sports goods, handloom fabrics, handloom machine, defence goods and aluminium articles whereas arrival through railway booking are petroleum, coal, firewood, logs, timber, oilcake and gram husk, cement, fertilizers, railway material, defence goods and miscellaneous.
2. Round figures of the values have been discussed in the running text, while their exact figures have been given in brackets.
3. Survey included commodities such as cotton textiles cloth, sports goods, cutlery, milk and milk-made items, sugar, rubber products, cotton textile yarn, transformers, surgical goods, agricultural implements, solder-wire, *gobar* gas equipments, liquor, pharmaceutical sugar cube and chemicals.
4. These 16 commodities have been referred in the preceding part of the section.

3

FLOW PATTERNS OF SELECTED COMMODITIES

INTRODUCTION

THIS CHAPTER is an extension of the preceding chapter in the form of case studies. It aims at analysing the raw material and production linkages, income and employment generation and distribution system of selected commodities. Selection of commodities was mainly based on the demand (local as well as non-local) of raw material and production; and the supply of the finished products (regional and national/international). Keeping these factors into consideration four commodities were selected for detailed analysis. These commodities are textile, sports goods, cutlery and dairy. While production of textile and sports goods mostly depends on the non-local raw material, production of cutlery and dairy depends upon the local raw material. In terms of supply, cutlery and dairy cater to the needs of local as well as regional demands whereas textile and sports goods cater to the national and international demands.

This exercise is based on the field-work carried out during 1986.* Field enquiry was based on the stratified random sample survey. Interviews were held based on the proportionate representation of

* The field survey was conducted in the months of May, August and November 1986.

small, medium and large units. The survey included the enquiry about raw material, labour, capital, technology, production, storage, supply and income generation, besides problems and prospects of development. Besides collection of primary data, secondary data were also collected to focus upon the various issues of these industries. Informations were collected from the offices of the General Manager, District Industries Centre, Meerut, Assistant Director of Industries, Meerut region, Labour Commissioner, Meerut, and Labour Unions of each industries. Data regarding production and supply relate to the period of 1984, whereas data regarding growth trends relate to the period 1973-84 in case of dairy, 1975-84 in case of textiles and 1981-84 in case of sports goods and cutlery industries. Data, thus, collected had a large variations, as such uniform line of explanation could not be adopted for all industrial units. However, an attempt has been made to make the observations comparable and coherent as far as possible. Statistical and cartographic methods have been used to substantiate the results. Explanations about the statistical exercises and maps have been given along with the case studies. Detailed discussion about each industry is given below:

Textiles

Textiles provide the largest employment of industrial labour in the country. It is the only industry that functions both in urban as well as in rural areas. India is the third largest producer of cotton textiles (first two being USSR and China) in the world.

Meerut is an important centre of cotton textiles in the north-western India. Despite series of challenges in the past, industry has not only recovered but strengthened its base for further development. According to an official estimate, there are about 70,000 weavers in the district.[1] This industry is mostly small scale and at household level, organised through chain of production units from Meerut down to the distant villages. Areas with this kind of industrial set-up and development are few in the country. Recent changes in new textile policy, financial aid and, above all, relatively advanced economic base of the region have accelerated the pace of cotton textile industry in Meerut. Meerut has a distinction of establishing strong space relations and experiencing the green as well as white revolutions in addition to being too close to Delhi -- the national Capital.

Due to developed infrastructure and economic base, Meerut region experienced fast urban-industrial growth. Clothing being one of the basic necessities of life, textile activity has emerged as the common industrial function in this region. Of the total workers (about 11.89 per cent) and of the total industrial units (about 9.05 per cent) are associated with the production of cotton textiles. In terms of the value of the product textile production, through Gandhi Ashram units in the region, was worth Rs. 5.73 crores and, through the units of handlooms and powerlooms, production in Meerut city was about Rs. 874 million during 1984-85. It is, thus, clear that cotton textile is the next most important economic activity after agriculture in Meerut region.

Data used in this exercise refer to the periods 1975, 1981 and 1984. It was obtained from the offices of the Assistant Director of Industries – Handlooms, Meerut, Handloom Intensive Development Project Ltd., Meerut, and the General Manager, District Industries Centre, Meerut.

Statistical and cartographic methods have been used to substantiate the results. Percentage, seasonal and monthly gaps in flows, functional concentration and service range of the centres have been worked out. Proportional circles, bar diagrams and graphs have been plotted to express the flow relations and hierarchic dependence over the space.

Place and time dimensions of textile flows and their price values have been worked out to analyse the extent of variation, association and influences. Coefficient of variation,[2] coefficient of correlation[3] and regression equations[4] have been worked out for the purpose.

The calculation of the functional hierarchy of textiles is based on the composite scores of the indices of functional concentration and the extent of areal influence exerted by a centre. The lesser the score value, the lower is the order of the centre. Indices of concentration include production, collection, printing, storage, sale and supply, whereas extent of areal influence includes area under functional dependence. Indices have been worked out by normalising values with their respective average values whereas composite scores have been worked out by adding the indices. Order of the centres has been determined with the help of the marked gap in the array of composite scores.

The case study is divided into two units of explanations. The first deals with the growth of the cotton textile activity in Meerut city and the second deals with the localisation of this industry, its functional dependence over the space.

Growth of Textiles

Meerut city has had textile as one of its main functions from the historical past. Moreover, textile has attained a prominent position recently. Unlike the planned growth of big mills in major textile centres, Meerut experienced textile activity as a natural consequence over time. Recently, for about a decade, growth of textile activity has been accelerated due to the migration of weavers from rural areas, financial support for the industry, availability of raw material, machines and, above all, due to growing domestic and export demands of textile products.

Meerut city recorded 178.8 million metres of textiles, worth Rs. 874 million, during 1984. There has been sharp rise both in terms of units of production and in absolute production. However, production per unit remained low. There were 10,580 handlooms and 4,536 powerlooms in the city in 1975. By 1984, number of handlooms rose to 15,000 and that of powerlooms 7,400 with the respective growth rates as 4.64 and 6.99 per cent (Fig. 3.1). Similarly, output of textile through handlooms was 63.48 million metres and 54.55 million metres was through powerlooms in 1975. Output increased by 51.49 per cent during 1975-1984. It is, therefore, clear that despite rise in the amount of production and the number of units, productivity remains to be low. It is apparent that units do not produce to their normal capacity. Major changes are needed to raise the level of textile activity and its productivity (Table 3.1).

TABLE 3.1: Growth of Cotton Textiles in Meerut City — Units and Production

Year	*No. of handlooms*	*Handloom-made cloth (in million metres)*	*No. of powerlooms*	*Powerloom-made cloth (in million metres)*
1975	10,580	63.48	4,546	54.55
1976	11,735	70.41	5,640	67.68
1977	12,940	77.64	5,780	69.36
1978	13,528	81.16	6,320	75.84
1979	14,000	84.00	6,500	78.00
1980	14,200	85.20	6,680	80.16
1981	14,320	85.92	6,850	82.20
1982	14,640	87.84	7,000	84.00
1983	14,720	88.32	7,200	86.40
1984	15,000	90.00	7,400	88.80

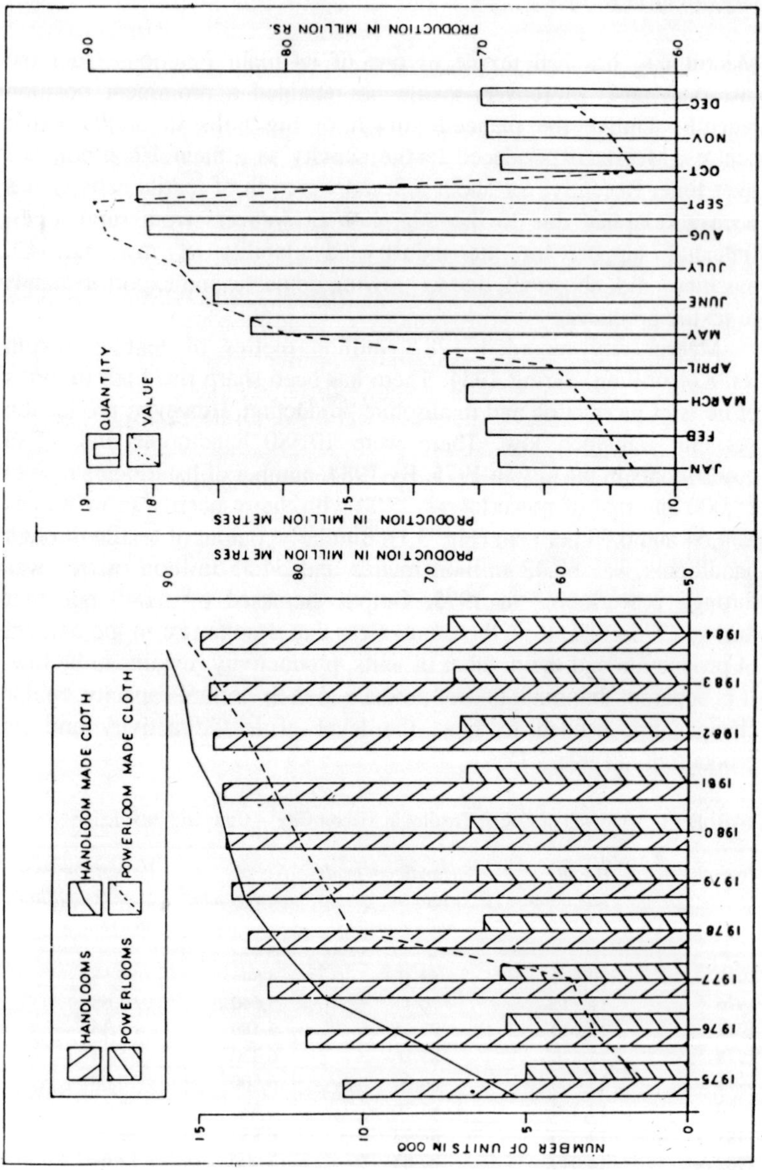

FIG. 3.1: Meerut city — Growth of cotton textiles (1975-84) and trends of monthly production — 1984.

Textiles and Gandhi Ashrams in Meerut Region

Gandhi Ashrams act as the organised system of textile production apart from handlooms and powerlooms sectors in Meerut region. The functions of Gandhi Ashram include spinning, weaving, dyeing and printing at the first level and purchase, sale and storage at the second level. Gandhi Ashrams are functioning at small and medium levels in rural as well as urban areas. These units provided a regular employment to about 1,000 (1,016) persons and casual employment to about 5 lakhs (4,90,048) persons during 1984-85. The regional annual textile production was worth around Rs. 57 million and that of yarn production worth Rs. 19 (Rs. 18.69) million. Purchase of Yarn amounted to Rs. 14 million and dyeing and printing works Rs. 1 (1.15) million. Number of centres and stores were 268 during 1984-85. Every village is covered by the centre or sub-centre functionary. Textile activity is, thus, organised at a large scale though textile output could not be geared up. Table 3.2 provides details about production, purchase, employment, number of centres and stores at the zonal levels.

TABLE 3.2: Meerut Region — Functional Position of Gandhi Ashrams (1984-85)

	Functional details	*Saharan-pur*	*Muzaffar-nagar*	*Meerut*	*Hapur*	*Buland-shahr*	*Total*
1.	Yarn purchase (in Rs.)	92,66,783	—	47,38,700	—	—	1,40,05,483
2.	Dyeing and Printing works (in Rs.)	—	—	11,50,000	—	—	11,50,000
3.	Yarn production (in Rs.)	—	1,07,27,478	—	48,97,779	30,70,820	1,86,96,077
4.	Village industries production (in Rs.)	3,79,545	1,58,098	11,50,000	1,92,670	1,58,257	—
5.	Textile production (in Rs.)	1,28,97,500	1,84,38,137	87,43,900	1,00,50,415	72,07,563	5,73,37,515
6.	Employees in nos.	343	68	302	168	135	1,016
7.	No. of workers	16,723	11,218	6,535	7,013	7,560	49,049
8.	No. of stores and centres	90	26	60	40	52	268

Source: Based on the records (1984-85) of Gandhi Ashram Zonal Offices in Meerut Region.

Flow Links and Space Relations

The extent of inflow and outflow indicates relative nodality of a place and helps in identifying gaps over the space and time. These gaps offer scopes for increasing the textile activities over these areas. Meerut city acts as a nodal centre for textile activity in this region. Details about inflow and outflow of textiles are given below:

Inflow of Textiles: Meerut city registered an annual inflow of about half (0.50) million metres of cotton textile worth about Rs. 2 (1.97) million (Table 3.3). There has been a steady decline of nearly 12 (11.93) per cent during 1981-82 and 25 per cent during 1982-83 in the amount of inflow. Decline in the value of textile inflow was relatively lesser, i.e., 7.41 and 19.06 per cent, respectively. The period 1983-84 observed tremendous increase of 30.57 per cent in the quantity and 37.20 per cent in the value terms. Inflow is mostly from Pilkhuwa (33.29%) followed by Kairana (15.46%) and Khatauli (11.25%). Among other centres Sardhana (9.70%), Shahpur (9.40%), Islamnagar (8.06%), rural areas of district Meerut (7.87%) and Manglore (4.97%) are important in terms of inflow to Meerut. The quantity of textile (170 million metres) outflow accounted for Rs. 835.45 million in 1984. Meerut also recorded supplies of cotton textiles from Sikandrabad, Deoband and Garhmukteswar before 1984.

TABLE 3.3: Textile Flow Relations of Meerut City — 1984

Places	*Inflow of textiles*				*Outflow of textiles*		
	Quantity (in metres)	*Percentage*	*Value (in Rs.)*	*Percentage*	*Places & States*	*Quantity (in million metres)*	*Percentage*
From rural areas of district Meerut	39,655	7.87	1,94,100.00	9.81	Kanpur	32.68	19.82
Sardhana	48,917	9.70	1,85,620.00	9.38	Punjab	35.76	21.68
Pilkhuwa	1,67,812	33.29	6,35,800.00	32.13	J & K	26.82	16.25
Islamnagar	40,632	8.06	1,78,330.50	9.01	Bihar	42.91	26.00
Manglore	25,082	4.97	90,676.00	4.58	M.P.	26.82	16.25
Shahpur	47,383	9.40	1,87,609.50	9.48			
Kairana	77,952	15.46	2,91,550.00	14.74			
Khatauli	56,708	11.25	2,14,923.50	10.87			
Total	5,04,141	100.00	19,78,609.50	100.00		165.00	100.0C

Monthly arrival reveals seasonal fluctuations (Table 3.4). It fluctuates from about one (0.96) per cent of the total annual arrival in December to about 25 per cent in April. About 47 per cent of the total textile annual arrival was recorded during the months of March and April. This period is classed as the peak arrival. Quantity of moderate arrival was recorded during the months of May to August; the share of arrival being 36 (35.87) per cent. Months of January, February and September registered low quantity of arrival sharing about 10 (10.16) per cent of the total. The arrival further falls down during November and December, the lean period, and accounted for merely three per cent (Fig. 3.2). Besides variations in the monthly arrivals, there are variations in the monthly supply of the centres. Islamnagar, for instance, registered regular supply for all the months (January to December), Pilkhuwa 11 months (January to November), Kairana 10 months (January to October), Khatauli and Sardhana nine months (January to April and July to October) and parts of district Meerut only seven months (March to September).

Outflow of Textiles: Cotton textiles present variation in its outflow over time and space. Outward supply of textiles rose from the level of about Rs. 6 (Rs. 5.94) crores in 1982 to Rs. 8 (8.35) crores in 1984, registering an annual growth rate of about six per cent. In terms of growth, there has been an increase of about 22 (21.89) per cent and four (4.42) per cent for the corresponding periods of 1982-83 and 1983-84 and a decline of 10 (9.74) per cent for the period of 1981-82. The period (1981-84), on the whole, registered an average increase of about 15 (14.88) per cent. Of the total outflow, States of Bihar accounted (Fig. 3.3) for 26 per cent, Punjab 22 (21.68) per cent, Uttar Pradesh (Kanpur)[5] about 20 (19.82) per cent, Jammu and Kashmir and Madhya Pradesh for 16 (16.25) per cent each. Monthly outflow (Table 3.4) presented two trends:

(i) The period of increasing outflow – May to September. A regular increase from 9.31 to 10.37 per cent was recorded for this period. During 1984, the average cloth supply was 16.36 per cent,

(ii) The period of constant outflow – October to April. This period observed a constant supply of about 112 (112.12) million metres of cloth (7.35 %).

Since there are marked seasonal variations in the weather conditions, demand of cotton textiles remains low during winter because of the high substitution by the woollen textiles. On the contrary, output and outflow of cotton textiles increase during summer following increased demand of the product due to hot weather conditions.

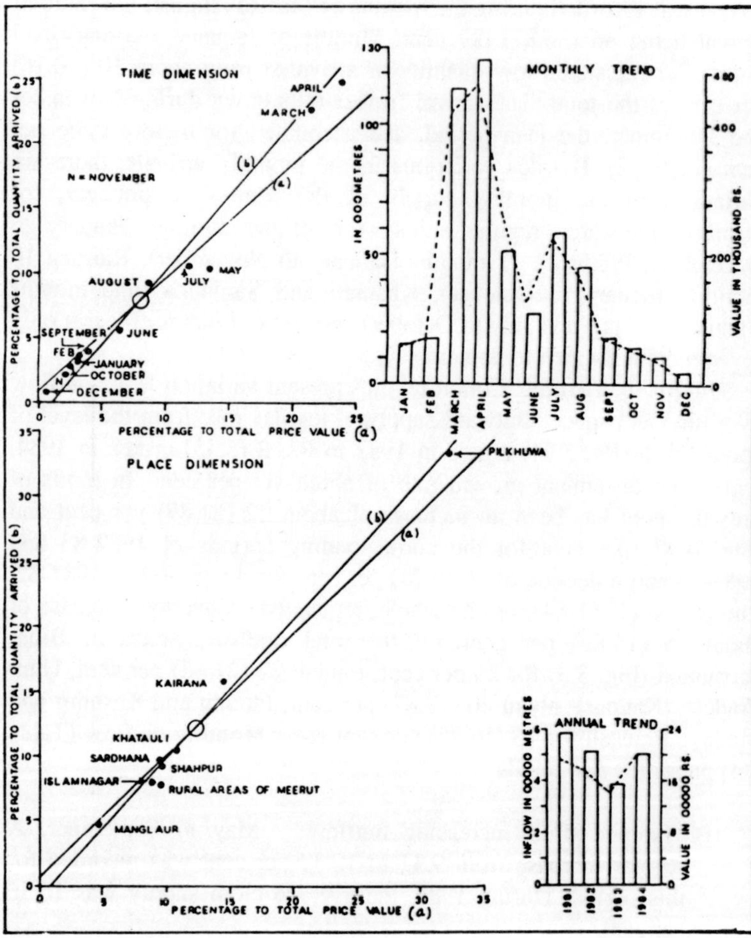

FIG. 3.2: Cotton textile arrival at Meerut — 1984.

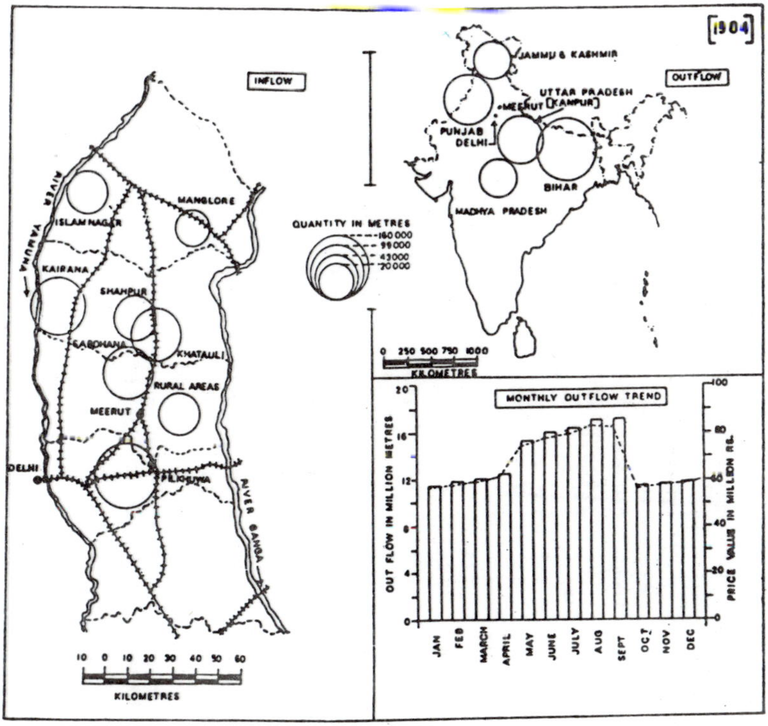

FIG. 3.3: Space relations of Meerut for cotton textiles — 1984.

Data further reveal dissimilar trends in the inflow and outflow of textiles which indicate the presence of missing links in the distribution. Since forward linkages from Meerut are well-organised and straight, outflow appears to be in ever-rising tendency. As against this, backward linkages are curved and unorganised. Due to missing links and areal gaps, arrival of textile is affected both over time and space. Production of cotton textile in Meerut city plays significant role in increasing the level of outflow.

TABLE 3.4: Meerut City — Monthly Behaviour of Production, Inflow and Outflow of Cotton Textiles in 1984

Months	*Production of Textiles*				*Inflow of textiles*				*Outflow of textiles*			
	Amount (in million mts)	*Percentage*	*Value (in million Rs.)*	*Percentage*	*Amount (in 000 mts)*	*Percentage*	*Value (in 000 Rs.)*	*Percentage*	*Amount (in million mts)*	*Percentage*	*Value (in million Rs.)*	*Percentage*
1	2	3	4	5	6	7	8	9	10	11	12	13
January	12.74	7.13	62.29	7.12	15.45	3.06	59.42	3.00	11.56	7.01	57.38	7.12
February	12.97	7.25	63.44	7.26	17.78	3.53	68.15	3.45	11.80	7.15	58.44	7.25
March	13.20	7.38	64.54	7.38	114.04	22.62	419.39	21.20	12.12	7.35	59.81	7.42
April	13.51	7.56	66.06	7.56	125.31	24.86	461.49	23.32	12.44	7.54	61.24	7.59
May	16.50	9.23	80.68	9.23	51.45	10.21	262.54	13.27	15.35	9.30	74.39	9.22
June	17.17	9.60	83.98	9.60	27.81	5.52	125.82	6.36	16.01	9.70	77.40	9.60
July	17.50	9.79	85.57	9.79	58.68	11.64	225.22	11.38	16.36	9.92	78.96	9.79
August	18.19	10.17	88.97	10.18	45.40	9.00	171.04	8.64	17.04	10.33	82.03	10.17
September	18.27	10.22	89.36	10.22	18.10	3.59	70.11	3.54	17.15	10.39	82.29	10.21
October	12.72	7.11	62.29	7.12	14.65	2.91	56.14	2.84	11.54	6.99	57.36	7.11
November	12.96	7.25	63.39	7.25	10.61	2.10	40.65	2.06	11.78	7.14	58.37	7.24
December	13.07	7.31	63.76	7.29	4.86	0.96	18.63	0.94	11.85	7.18	58.69	7.28
Total	178.80	100.00	874.33	100.00	504.14	100.00	1,978.60	100.00	165.00	100.00	806.36	100.00

Responses of Textile Flows

Results reveal that textiles flows (inflow as well as outflow) vary significantly from place to place; the Coefficient of Variation (C.V.) being 66.8 per cent. Textile prices have relatively lesser variations (C.V. – 62.85) over various places. The relationship between the two attributes is highly positive, though it is not exactly 'one is to one'. Regression coefficients reveal that per unit change in the textile flow (over places) affects prices by 0.93 unit. Similarly per unit change in price influences textile flow by 1.05 unit. It is, therefore, important to note that while textile flow tends to decrease with distance, its value tends to increase. (Time dimension presents slightly different result to that of place dimension). The coefficient of variation for textile flow was 91 (91.34) and for price value 87 (86.78) per cent. The two attributes (over time) present high degree of positive correspondence. Regression coefficients reveal that per unit change in price influences textile flow by 1.04 unit. Similarly, per unit change in textile flow affects price by 0.94 unit level (Table 3.5).

TABLE 3.5: Textile Flow and Price Value Response

	Dimension of Distribution	*Place dimension*	*Time dimension*
1.	*Standard Deviation* :		
	a) Textile flow	8.35	7.59
	b) Price value	7.85	7.22
2.	*Coefficient of Variation* :		
	a) Textile flow	66.80	99.34
	b) Price value	62.85	86.78
3.	*Coefficient of Correlation* :	+0.996	+0.992
4.	*Regression Equation* :		
	a) Textile flow on price value	*a*=0.937b+0.781	*a*=0.944b+0.464
	b) Price value on textile flow	*b*=1.059a-0.736	*b*=1.043a-0.360

It is, thus, established that textile flow presents high degree of variation association but less than the normal influence both over time and place, though it is more with time than with the place. Price, on the other hand, presents less variation and more than the normal influence over place and time. As such price has influential role to accelerate the flow over place and time. It increases both ways, i.e., with the extent of flows and with the distance. It is possible to have still larger impacts over textile flow if price fluctuations are controlled.

Hierarchy of Textile Centres

Interaction for textile raw material, production, consumption and sale has assumed significance and, as such, forms the basis for the identification of various orders of textile centres. While raw material, production, stock and inflow form the basis for hierarchic dependence among the centres, outflow forms the basis for the identification of linkages with the centres of national and international importance. For textile production, collection, repair, sale and technical aid, Meerut exercises its hierarchic and nodal influence upon its region. It includes districts of Saharanpur, Muzaffarnagar, Meerut, Ghaziabad and Bulandshahr.

Results reveal (Fig. 3.4) three orders of textile centres in Meerut region, viz., primate centres, intermediate centres and the production centres. Explanation about each type of the centre is given below:

Primate Centres : There are two such centres in the region – Pilkhuwa and Meerut – with their respective score values as 4.5819 and 3.7689. Pilkhuwa has 120 medium scale factories, 1,400 powerlooms and three *niwar* factories. As per an estimate there are 7,000 persons employed in textile production, 2,000 in washing cloth sheets, 100 in *kundis*, 500 in pressing machines, 500 in dyeing, printing, screening and knot making and remaining about 200 in stamping and designs. As mentioned earlier, Pilkhuwa is a regular supplier of handloom cloth to Meerut besides its supplies to other centres in the country and abroad. Meerut, another primate centre, has the facility of weavers' service centre, Handloom Intensive Development Project, Zonal Centre of Gandhi Ashram, Yarn Supply Centre, Assistant Director of Industries (Handlooms), Meerut Division, and District Industries Centre. Meerut, being bigger Urban Centre, has 52,000 persons (32,253 weaver only) in about 15,000 handlooms and 7,400 powerlooms, besides one medium scale spinning and weaving mill in co-operative sector. The centre recorded an annual production of 178.8 million metres of cloth. There are about 500 traders and 80 printing centres. It provides the facility of cloth collection, supply of raw yarn, storage and outward supply.

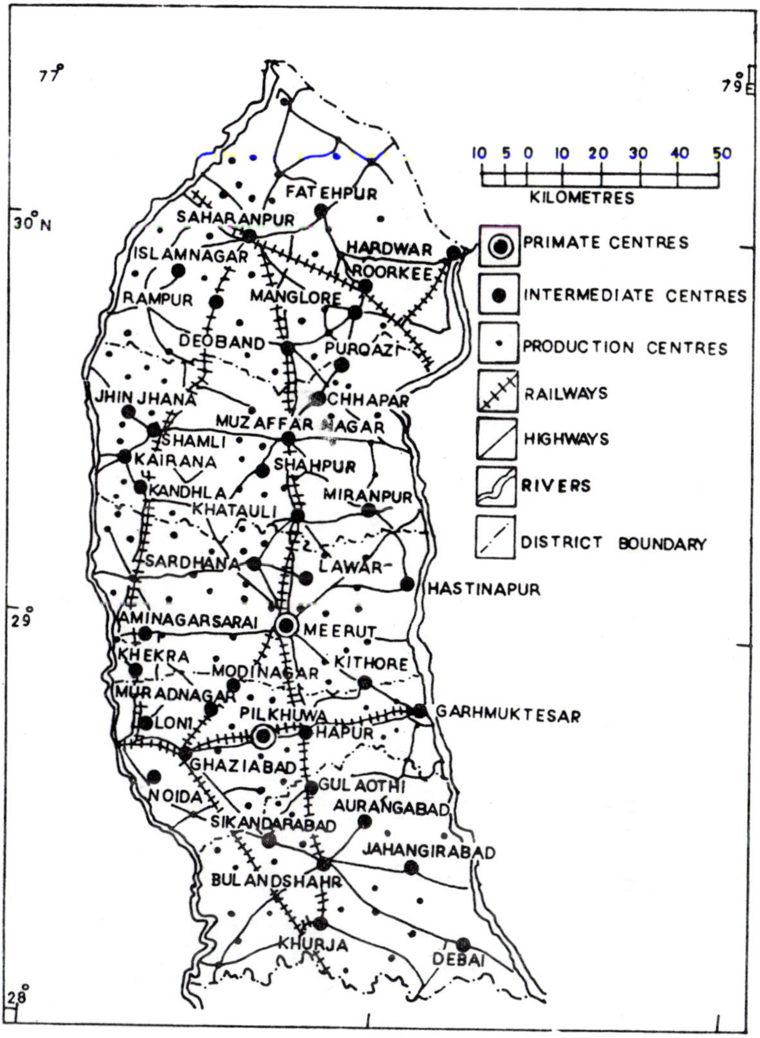

FIG. 3.4: Hierarchy of textile centres in Meerut region.

Intermediate Centres: There are 35 such centres in the region. The composite scores range from 0.531 to 2.391. These centres act as the link between production and for the sale of cloth, purchase of yarn, repair of machinery, storage and printing; they act as points of textile collection and yarn supply to the neighbouring production units. On an average, one intermediate centre looks after seven production centres

TABLE 3.6: Hierarchy of Textile Centres in Meerut Region

Primate centres	*Composite scores*	*Intermediate centres*	*Composite scores*	*Intermediate centres*	*Composite scores*	*Districts*	*No. of production units*
1	2	3	4	5	6	7	8
Pilkhuwa	4.5819	Fatehpur	0.651	Lawar	1.896	Saharanpur	87
Meerut	3.7689	Saharanpur	2.153	Sardhana	2.391	Muzaffarnagar	19
		Islamnagar	1.621	Hastinapur	2.035	Meerut	56
		Rampur	1.371	Khekra	1.426	Ghaziabad	32
		Roorkee	1.876	Kithore	1.639	Bulandshahr	45
		Hardwar	1.636	Aminagar Sarai	1.302		
		Deoband	1.961	Hapur	2.021		
		Purqazi	0.653	Garhmukteswar	1.628		
		Chhapar	0.531	Modinagar	1.489		
		Muzaffarnagar	2.051	Loni	0.671		
		Jhinjhana	0.616	Noida	0.650		
		Kairana	1.231	Gulaothi	1.651		
		Kandhla	1.621	Sikandrabad	1.873		
		Shamli	1.339	Aurangabad	1.421		
		Shahpur	1.891	Bulandshahr	2.061		
		Khatauli	1.971	Debai	0.631		
		Miranpur	0.625	Khurja	0.760		
				Jahangirabad	1.029		

for their textile demand and collection. Some of the important centres in this level of hierarchy are Sardhana, Hastinapur, Saharanpur, Bulandshahr, Muzaffarnagar, Hapur, Deoband, Khatauli, Lawar, Sikandrabad, Islamnagar and Shahpur (Table 3.6).

Production Centres: These centres act as the production points at the lowest level of hierarchy. Higher order functions such as collection, storage and trade have not developed in these centres. These centres depend upon the intermediate and primate centres for higher order functions in textiles. There are 238 production centres in Meerut region. They function under private sector, Gandhi Ashram and Sadhan Sahkari Samitis. The composite scores of these centres are less than 0.50. These centres have a few units of production, about 10 to 60 weavers and about 40 to 200 women spinners.

The foregoing analysis leads to the following conclusions:

(i) Cotton textile industry has developed as one of the chief industrial activities of Meerut region particularly of Meerut city. Being widespread economic activity, it provides maximum employment (about 12 per cent) after agriculture. Textile industry suffers from uncoordinated development between rural and urban units. It has led to the large scale out migration of weavers from rural areas. In the course of city-ward migration of weavers, textile dependency has increased significantly whereas labour productivity remained low,

(ii) Despite low productivity, large scale employment in textile units has raised the absolute production. Cotton textile industry is, as such, characterised by high production, high arrival and high outflow. Meerut city, with its administrative, infrastructural and functional nodality, acts as the primate centre of textile production and maintains regular flow links with its region and centres in northern and central India. Price fluctuations and missing functional links, at various levels of hierarchy, have adversely affected the textile flow, and

(iii) Fairly large number of production centres (233) have come up in the passage of time as a result of historic continuity. Textile centres have more regular distribution in the region. Meerut city has established relations with the production centres in the region. It acts as a regional node for the collection and distribution of textile products.

Sports Goods

One of the prime objectives of human resource development in the country is to encourage sports activity at all levels. With the result, sports activities have become part of the socio-educational milieu of India. Subsequently, demand for sports goods has risen significantly for the past few years which has resulted into the growth of sports goods industry in the country. Meerut city is one of the chief centres of sports goods industry in the country. This industry has recently gained international significance with the establishment of Process-cum-Product Development Centre here under the joint assistance of Government of India and U.N.D.P. in 1984. A sum of Rs. 90 lakhs was provided by the Central Government of India, Rs. 20 lakhs, besides land and workshop, by the State Government of U.P. and U.S. $ 5.88 lakhs by U.N.D.P. as a financial support to the complex. The plan included purchase of local machines worth about Rs. 5.15 million and imported machines worth U.S. $ 1.47 lakhs. Spatial distribution of sports goods industries in Meerut is along Surajkund road, Victoria Park and Lal Kurti areas. Some establishments have also started functioning in new sports goods complex on Meerut-Delhi highway. This industry has also developed in a few villages near Meerut city.

Study of sports goods industry includes factors such as number, type, production and employment on the one hand and growth trends and flow relations on the other. Industrial units were classified based on the type of raw material used, capital investment and employment. Finally, production values were compared, problems identified and financial assistance and policy support were explained along with the text. Since most of data have been collected through field survey, temporal comparisons were not possible. Based on the nature of data availability, simple statistical techniques such as percentage, total values, range and variations have been applied to explain the results. Although advanced statistical measures could not be used in the study, utmost care has been taken to draw conclusions free from any bias.

Production Units and Employment

Meerut city has about 1,000 industrial units of sports goods industry with the estimated employment to about 15,000 workers.[6] Nearly 87 per cent of the workers belong to the category of unskilled workers with poor economic background. Of the total workers, 40 per cent are the child workers, 50 per cent female workers and remaining are the adult male workers. This is a labour intensive industry and provides an average employment to about 30 full and part time workers per unit. There are only 307 industrial units in the city which are registered with the department of industries under Small Scale Industries (SSI). Similarly, only about 2,000 (1,746) workers are registered with the labour office, Meerut.

Investment varies significantly from unit to unit. There are as many as 262 units with low investment (less than Rs. 10,000), 35 with medium investment (between Rs. 25,000 to Rs. 1 lakh) and 10 with the high investment (more than Rs. 1 lakh). Further variations have been recorded at each level of industrial units. For instance, category of low investment has 42 units having investment between Rs. 15,000 to 25,000, 55 units between Rs. 10,000 to 15,000 and remaining 165 units with investment below Rs. 10,000. Similarly category of medium investment has 22 units having investment between Rs. 50,000 to 100,000 and 13 units between Rs. 25,000 to 50,000. In terms of production, 60 units (Table 3.7) produce inflatable balls with bladders followed by 57 units producing cricket bats and 53 units producing racquets – badminton, tennis and squash. Production of cricket/hockey balls and shuttle-cocks is carried out by 45 units. It is followed by the production of gymnastic/athletic equipments by 35 units. T.T. tables and stumps are produced by 19 units, also the same number of units produce sports nets and sports shoes. While rubber and plastic goods are produced by 15 units, base bats and badminton presses are produced by eight units and hockey sticks by six units. It is worth mentioning here that certain units (45) deal with the production of more than two items. Variety of productions by these units have made it complex to categorise them item-wise. Classification presented above is a simplest form of description for sports goods. Meerut city produces, on an average, 128 dozens of tennicoit rings, 1,080 dozen rubber balls and 2,650 dozen bladders per day. Since the information about the daily output of other items was not available, monthly production was analysed in the text. Monthly production of sports goods presents marked variations. November to February is the period of peak production whereas May-June period is of lean production. Nearly 60 per cent of the annual production is recorded during the peak period and remaining 40 per cent of the production is carried out over rest of the eight months.

TABLE 3.7: Meerut City — Sports Goods Units and Their Production — 1984

Items of production	*Unit of production*	*Production per day*	*No. of units*
1. *Rubber/plastic-based sports goods*			
(i) Bladders	Dozen	2,650	15
(ii) Rubber/compo balls	Dozen	1,080	11
(iii) Tennicoit rings	Dozen	128	7
2. Soft leather goods (gloves, leg guards, etc.)	Dozen	—	16
3. Gymnastic/athletic equipments	Dozen	—	14
4. *Cricket bats*			
(i) Kashmir willow	Dozen	32	11
(ii) White wood	Dozen	125	46
5. *Badminton/tennis/squash racquet*			
(i) Badminton	Dozen	182.5	46
(ii) Tennis	Dozen	4	5
(iii) Squash	Dozen	2	2
6. Cricket/hockey balls	Dozen	316	23
7. Shuttle cocks	Dozen	721	22
8. Stumps	Dozen	39.5	11
9. T.T. bats	Dozen	17	5
10. Badminton Presses	Dozen	15	3
11. Carrom-boards	Pieces	413	15
12. Anklets/knee caps	Pieces	—	6
13. Hockey sticks	Pieces	—	6
14. T.T. tables	No.	16	3
15. Base bats	No.	10	5
16. Sports nets	No.	—	15
17. Sports shoes	No.	--	4
18. Inflatable balls (with bladders)	No.	1,647	60

Production

On the basis of raw material used in the production, sports goods may be classified as leather-based, wood-based and rubber-based. Leather-based products formed 33 per cent, wood-based 26.50 per cent and rubber-based 23.70 per cent of the total production. During 1984, the production of sports goods was worth Rs. 73.48 million weighing about 1,854.63 tonnes (Table 3.8). Amount of sports goods production varied significantly. Production (in value terms) ranged between Rs. 20,000 in case of base bats to Rs. 17 (17.34) million (Table 3.9) in case of inflatable balls. The level of very high production was recorded in cases of inflatable balls and soft leather goods, the respective amounts

TABLE 3.8: Meerut City — Itemwise Trends of Production in Sports Goods Industry

(in weight terms)

Items	Production (in tonnes)								Average annual growth rate 1981-84
	1981	*Percent-age*	*1982*	*Percent-age*	*1983*	*Percent-age*	*1984*	*Percen-tage*	
1. Soft Leather goods	190.09	16.00	198.16	14.01	224.92	13.79	233.00	12.56	+ 7.52
2. Inflatable balls	117.93	9.93	113.34	8.01	111.82	6.86	110.93	5.98	— 1.98
3. Cricket bats	58.64	4.94	62.51	4.42	65.90	4.04	70.59	3.81	+ 6.79
4. Athletics/gymnastic equipment	329.60	27.74	373.80	26.43	428.23	26.26	490.71	26.46	+ 16.29
5. Carrom-boards	85.33	7.18	219.20	15.50	333.33	20.44	454.82	24.52	+144.34
6. Sports hosiery	15.00	1.26	16.87	1.19	17.50	1.07	20.00	1.08	+ 11.11
7. Sports wear	0.96	0.08	0.92	0.07	0.96	0.06	1.00	0.05	+ 1.39
8. Bladders	35.88	3.02	42.36	2.99	43.29	2.65	48.76	2.63	+ 11.96
9. E.P.N.S. sports wear	125.00	10.52	128.00	9.05	129.80	7.96	130.00	7.01	+ 1.33
10. Sports nets	20.64	1.74	21.99	1.55	23.87	1.46	25.99	1.40	+ 8.64
11. Rubber and plastic sports goods	79.20	6.67	98.95	7.00	100.90	6.19	104.90	5.66	+ 10.82
12. Hockey sticks	11.95	1.01	11.06	0.78	10.55	0.65	10.89	0.59	— 2.96
13. Racquets	12.38	1.04	13.08	0.92	14.14	0.87	15.48	0.84	+ 8.35
14. Shuttle-cocks	6.08	0.51	5.96	0.42	6.08	0.37	6.57	0.35	+ 2.68
15. Sports shoes	8.33	0.70	9.58	0.68	10.30	0.63	11.58	0.62	+ 13.00
16. T.T. tables	11.20	0.94	13.06	0.92	17.73	1.09	25.20	1.36	+ 41.66
17. T.T. bats	2.86	0.24	4.96	0.35	4.55	0.28	4.00	0.22	+ 13.28
18. Cricket/hockey balls	66.92	5.63	70.37	4.98	75.86	4.65	77.91	4.20	+ 5.47
19. Stumps	8.95	0.75	9.00	0.64	9.75	0.60	10.10	0.54	+ 4.28
20. Base balls	1.00	0.08	1.10	0.08	1.16	0.07	1.20	0.07	+ 6.66
21. Base bats	0.20	0.02	0.15	0.01	0.20	0.01	1.00	0.05	+133.33
Total	1188.14	100.00	1414.42	100.00	1630.84	100.00	1854.63	100.00	+ 18.69

Source: Sports Goods Export Promotion Council, New Delhi - 110 055.

TABLE 3.9: Meerut City — Item-wise Trends of Production in Sports Goods Industry

Item	*Production (in lakh rupees)*							
	1981	*Percentage*	*1982*	*Percentage*	*1983*	*Percentage*	*1984*	*Percentage*
1. Soft leather goods	97.33	15.97	101.46	15.67	115.16	16.68	120.26	16.37
2. Inflatable balls	182.72	29.99	175.60	27.12	173.25	25.09	174.33	23.72
3. Cricket bats	73.30	12.03	78.14	12.07	82.38	11.93	88.24	12.01
4. Athlctics/gymnastic equipments	36.69	6.02	41.61	6.43	47.67	6.90	54.65	7.44
5. Carrom-Boards	8.00	1.31	20.55	3.17	31.25	4.53	42.64	5.80
6. Sports hosiery	6.00	0.984	6.75	1.04	7.00	1.01	8.00	1.09
7. Sports wear	0.24	0.04	0.23	0.04	0.24	0.034	0.25	0.03
8. Bladders	15.95	2.62	18.83	2.91	19.24	2.79	21.67	2.95
9. E.P.N.S. sports wear	12.50	2.05	12.80	1.98	12.98	1.88	13.00	1.77
10. Sports nets	7.78	1.28	8.29	1.28	9.00	1.30	9.80	1.33
11. Rubber and plastic sports goods	15.84	2.60	19.79	3.06	20.18	2.92	20.98	2.86
12. Hockey sticks	11.95	1.96	11.06	1.71	10.55	1.53	10.89	1.48
13. Racquets	28.13	4.62	29.72	4.59	32.14	4.65	35.18	4.79
14. Shuttle-cocks	24.34	3.99	23.84	3.68	24.35	3.53	26.28	3.58
15. Sports shoes	5.00	0.82	5.75	0.89	6.18	0.89	6.95	0.95
16. T.T. tables	2.40	0.39	2.80	0.43	3.80	0.55	5.40	0.73
17. T.T. bats	7.16	1.18	12.40	1.915	11.38	1.65	10.00	1.36
18. Cricket/hockey balls	71.70	11.77	75.40	11.65	81.28	11.77	83.48	11.36
19. Stumps	1.79	0.29	1.80	0.28	1.95	0.28	2.02	0.27
20. Base balls	0.50	0.08	0.55	0.08	0.58	0.08	0.60	0.08
21. Base bats	0.04	0.006	0.03	0.005	0.04	0.006	0.20	0.03
Total	609.36	100.00	647.40	100.00	690.60	100.00	734.82	100.00

Source: Sports Good Export Promotion Council, New Delhi - 110 055.

being Rs. 17 (17.43) million and Rs. 12 (12.02) million. The two types of products together accounted for 40 per cent (Rs. 29.45 million) of the value and 19 per cent (343.93 tonnes) of the weight. Similarly, the level of high production, with a value worth Rs. 17 (17.17) million and weight worth 148 (148.5) tonnes, was recorded in cases of cricket bats and cricket and hockey balls. The share of these items in the total production was 23 (23.36) per cent in terms of value and 8 per cent in terms of weight. Of 21 items of sports goods production, six items recorded medium production. The value of the production ranged between Rs. 2 (2.09) million in rubber and plastic goods to Rs. 5.46 million in athletic and gymnastic equipments. In addition to these two products, category of medium production includes carrom-boards, racquets, shuttle-cocks and bladders. The category shared about 27 (27.40) per cent of the total production in money terms and 60 (60.45) per cent in terms of weight. The value of the low level of production ranged between Rs. 5 (5.4) lakh in case of T.T. tables to Rs. 1 (1.3) million in case of E.P.N.S. Sports wear. The products listed in this category are E.P.N.S. sports wear, hockey sticks, T.T. bats, sports nets, sports hosiery, sports shoes and T.T. tables. These items accounted for about nine (8.71) per cent of the value and 12 (12.27) per cent of the weight. On the contrary, very low level of production was recorded in items, like, stumps, base balls, sports wear and base bats. In terms of value, production ranged between Rs. 20,000 in case of base bats to Rs. 2 (2.02) lakh in case of stumps. Share of this category in the total production was below one per cent, i.e., 0.41 per cent in terms of value and 0.71 per cent in terms of weight.

Growth Trends

Production of sports goods during 1975-76 was of the order of Rs. 31 (31.4) million which rose to the level of Rs. 112 (111.5) million by 1985-86 (Table 3.10). In terms of weight, Meerut city recorded about 1,200 (1,188.14) tonnes of sports goods production during 1981 which increased to the level of 1,900 (1,854.63) tonnes by 1984. Growth rate of sports goods ranged from one (1.39) per cent in case of sports wear to 144.34 per cent in case of carrom-boards with average rate for the industry as a whole being 19 (18.69) per cent. High to very high growth rate (from 41.66 to 144.34 per cent) was recorded in products, like, carrom-boards, base bats and T.T. tables. Medium growth rate (11.11% to 16.29%) was recorded in T.T. bats, sports shoes, athletic/gymnastic equipments and bladders and low growth rate (less than 9% per annum) was recorded in as many as remaining 12 products. Growth rates for the past four years (1981-84) revealed four trends of production, viz.:

(i) Declining trend of production,
(ii) Initial rise and subsequent declining trend of production,
(iii) Initial decline and subsequent increasing trend of production, and
(iv) Increasing trend of production.

Inflatable balls and hockey sticks fall in the first trend of production; T.T. bats in the second; sports wear, shuttle-cocks and base bats in the third and remaining 15 products fall in the fourth trend.

TABLE 3.10: Meerut City — Growth in Sports Goods Production in Registered Units (1975-86)

Year	*Total Production (in crore Rs.)*	*Foreign Export (in crore Rs.)*
1975-76	3.14	1.51
1976-77	3.70	1.67
1977-78	4.18	1.98
1978-79	5.57	2.30
1979-80	7.69	4.00
1980-81	10.00	7.00
1981-82	11.00	3.86
1982-83	10.50	3.31
1983-84	11.00	3.27
1984-85	11.00	3.86
1985-86	11.15	—

Source: District Industries Centre, Meerut City.

Thus, both in terms of production as well as growth rate, this industry is undergoing fast changes. Of the 21 products as many as 13 have registered an output worth over Rs. 1 million. Due to sharp rise in production and demand, sports goods industry has earned distinction among economic activities of Meerut city. Due to recent changes in industrial policy supported by liberal financial aid, sports goods industry witnessed high rate of production.

Flow Relations

Meerut has established strong forward as well as backward linkages in case of sports goods. Fast growth in the production has led to the greater dependence on areas supplying raw material and the demand of distant places for finished products. Consumption of sports goods in India was merely two per cent of the world market (Rs. 1230 crores) during 1984. Seventh Plan aimed at increasing the annual export of sports goods by Rs. 100 crores. Meerut, being the chief centre of sports goods production, has to increase its production for export. Before analysing the trade links of Meerut, it is necessary to give a brief account of the raw materials' links for the industry.

Spatial distribution of raw material reveals that Meerut receives leather from the neighbouring villages of Fatherpur, Shobhapur and Bhatipura and from cities of Madras, Kanpur and Hapur. Receipt of wood for the industry is received from Hastinapur, Nainital, Gonda and Gorakhpur. While cane is mostly received from Karnataka, Kashmir willow is received from Jammu. Supply of rubber is mainly from the States of Karnataka and Kerala. It is, thus, clear that bulk of raw material comes from distant places. These links need to be made cheaper, direct and regular in order to achieve the goal of higher production.

Products of sports goods industry are supplied to almost all major cities in India from where it gets re-distributed to the neighbouring towns and villages. While supply of these goods, within the country, was directed to the cities of Calcutta, Delhi, Madras, Bombay, Bangalore, Ahmedabad, Cochin and Kanpur; supply of these goods outside the country was mostly directed to European countries, USA, Middle-East countries, South and South-East Asian countries, USSR and Australia. Meerut city registered about 1,855 tonnes of weight as production of sports goods during 1984. Besides production, city also had a stock worth 700 tonnes from the previous years. Outflow of sports goods during 1984 was nearly 2,500 tonnes. Of the total outflow, about 78 per cent moved through highways and rest (about 22 per cent) through railways. Athletics/gymnastic equipments accounted for about 51 per cent of the total outflow. Soft leather goods, inflatable balls, rubber and plastic goods accounted for about 30 per cent. Cricket and hockey balls, cricket bats, bladders, sports nets, T.T. tables and sports hosiery accounted for about 18 per cent of the outflow whereas remaining items accounted for less than one per cent. Foreign export of sports goods was mainly through 151 units and was worth Rs. 3 (3.27) crores in 1984. Annual earnings from sports goods range between Rs. 7 to 8 crores. Meerut city is, thus, a leading centre of sports goods industry in the country. To meet the growing demands of home as well as international market, significant improvements (qualitative as well as quantitative) are needed to increase the present level of output. Consumption of sports goods in the world market was estimated as Rs. 1230 crores in 1984.

During the course of fieldwork following problems were noted:

(i) There is an acute need of a raw material bank to cater to the needs of sports goods industry in Meerut city. The fact that industry is going to expand in three production complexes, it is desirable to have a raw material bank to facilitate the functioning of production units round the year,

(ii) It was revealed, during the field survey, that rates of the products in the home market are much lower compared to rates in other producing countries. Efforts must be made to increase the earning of the producer through price policy, and

(iii) U.P. Forest Corporation, U.P., supplies a number of wood species for the industry. However, there remains a supply gap between the demand and the applications being processed and forwarded by D.I.C., Meerut, to the various depots of the Forest Corporation of U.P. Chemical treatment of wood is essential soon after the felling of the tree, so that it remains unaffected by fungus and black patches until it is sold to the manufacturer.

Cutlery Industry

Cutlery industry came into existence in Meerut around first half of the 19th century. Present industry is the result of a hereditary profession carried out over generations. Despite serious challenges in the past, industry survived due to established credit of the product in the consumer market. Cutlery is largely a household industry in Meerut and is found mostly in residential buildings. While front side of the house is used for manufacturing and selling of the cutlery product, backside is used for residential purposes. Industrial clusters are found in localities, like, Paramal, Sabungran, Kotla Street, Guzari bazar, Saddiq nagar, Lisari gate, Bhumian bridge, Baghpat gate, Islamabad,

Sohrab gate and Sarai Behleem. Besides Meerut city, centres like Parikshhatgarh, Mawana, Baragaon, Bhainsa, Naglasahu, Pohali and Lohia also have (cutlery) manufacturing units. Products of these places are brought to the Meerut city for storage and sale.

Results are based on the door-to-door survey in the manufacturing-cum-sale units of cutlery industry in Meerut city. Since industry is mostly unorganised, no official records of production, income, employment and flow relations were available. The text presented here is the response of the workers, owners and union secretary of the Scissors and Knife Manufacturers Association, Meerut. Due to the limitation of data, advanced statistical treatment of the exercise could not be done. However, attempt has been made to explain the profile of production, problems and prospects. Separate accounts for raw material, production, problems and suggestions are given below:

Raw Material and Industrial Set-up

Iron, brass, lime and coal are the chief raw materials used in this industry. Harness and tumburing furnaces, nuts, bolts and sharpening tools are the other necessary equipments of this industry. For most of its raw material demand, Meerut depends upon its local markets particularly Kabari bazar. Smaller processing units are engaged in the cutting of iron bars, hammering and sharpening of scissors and knives, whereas bigger ones are engaged in the nickel plating, assembly of parts, oiling, grading of the products, wrapping and packing. Products of the industry are scissors, knife, nail-cutter and locks of different sizes. There are about 350 units engaged in the manufacturing of scissors. Nearly 50 units are registered with the Department of Industries under Small Scale Industry (SSI). Of these about 25 units are engaged in the production of scissors along with the production of knife, locks, nail-cutters, etc. Most of the employees in this industry are Muslims. In all, there are about 15,000 part and full time workers. Most of the work is being done by adult male workers, however, child workers are also in significant (about 1,000) number. About 600 female workers get part time job in finishing works of this industry. Of these, only about 300 workers are registered with the labour office at Meerut. It is estimated that about 25,000 part/full time workers are engaged in this industry in the district as a whole.

Production

There are seasonal variations in the production of scissors. July-August and March-April are the periods of low production whereas rest of the eight months are the periods of normal production. While July-August recorded low production due to rainy season when rusting of the products becomes a common problem. Besides rusting, demand also decreases considerably during this period. March-April register low production due to temporary shift of labour from cutlery to crop cutting.

Output of cutlery industry shows marginal increase in its production over different years. During 1981 cutlery goods weighing about 73 tonnes were supplied to different parts of the country. Over a period of a year supply decreased to the level of 69 tonnes. This decline was largely due to the communal violence in the city which affected the cutlery production as well as its sale. There has been steady rise in the production after 1982. The amount of production was about 85 (85.38) tonnes during 1983 and about 87 tonnes during 1984. In terms of value, the output was worth Rs. 6 (6.1) million. The proportionate share of production between registered and unregistered units was 35 and 65 per cent respectively. Registered units accounted for the production of Rs. 1 (1.3) million whereas unregistered units for 5 (4.8) million. Productivity per worker as well as per unit remains low because the traditional to semi-modern technology, storage of raw material, power and capital.

Problems

Cutlery industry is largely in unorganised sector and is, therefore, faced with numerous problems of storage, quality control and marketing. Because of these bottlenecks, this industry continues to remain on the traditional technology and yields low return. Modernisation of the industry and its efficient organisation can help to boost the production.

There is an urgent need of cutlery industrial complex with the facility of a nickel plant, harness furnaces, tumburing furnaces, coal and lime depots. Due to non-existence of a corporation, a good amount of product awaits sale and gets spoiled within a short time.

Despite problems faced by the industry, it nas managed to maintain the level of its production (about Rs. 70,000 per unit). Necessary technical and financial support is essential for the rapid growth of this industry.

Milk

This exercise deals with the flow links of a dairy unit and its spatial and temporal trends. Having experienced the two successive revolutions (green as well as white) in the field of primary productions, regional economy has improved considerably. It has resulted into large scale surplus generation and increased purchasing power of the people. As a result several agro-based processing units have developed for the past few years. Feeder Balancing Dairy, Partapur, is the most important milk processing industrial unit in Meerut. This unit is in the co-operative sector. The dairy unit started processing of milk from Meerut and neighbouring districts. Both statistical as well as cartographic methods have been used to substantiate the results. Simple statistical methods, such as, growth rates, percentages and variations, have been used in the text. Like statistical, cartographic methods, such as, proportional circles, flow and bar diagrams, have been used to explain the spatial as well as temporal trends of arrival and outflow of milk and milk-made products.

Milk arrival at Feeder Balancing Dairy, Partapur, Meerut

Feeder Balancing Dairy, Partapur, registered an annual milk collection (Table 3.11) of about 21,000 tonnes of milk during 1984. Of the total milk arrival, nearly 63 per cent came from the rural areas of district Meerut. It is followed by about 4,000 tonnes of milk arrival from Gulaothi (district Bulandshahr) and about 2,000 tonnes from Masoori (district Ghaziabad). About 1,130 tonnes of milk was reported to have arrived from the rural areas of district Muzaffarnagar. It is important to note that Meerut recorded arrival of milk from far off places, like, Dehradun, Agra and Moradabad. Thus, Partapur dairy has established its inflow relations with areas lying within Dehradun, Agra and Moradabad triangle (Fig. 3.5).

The temporal flow of milk suggests an overall rise in the quantity of milk arrival over the years. While annual milk collection was only 2,800 tonnes in 1973, it attained the level of about 21,000 tonnes by 1984. There was a regular increase in the milk arrival during 1973-77, 1978-80 and 1983-84. On the contrary, a nominal fall in the milk

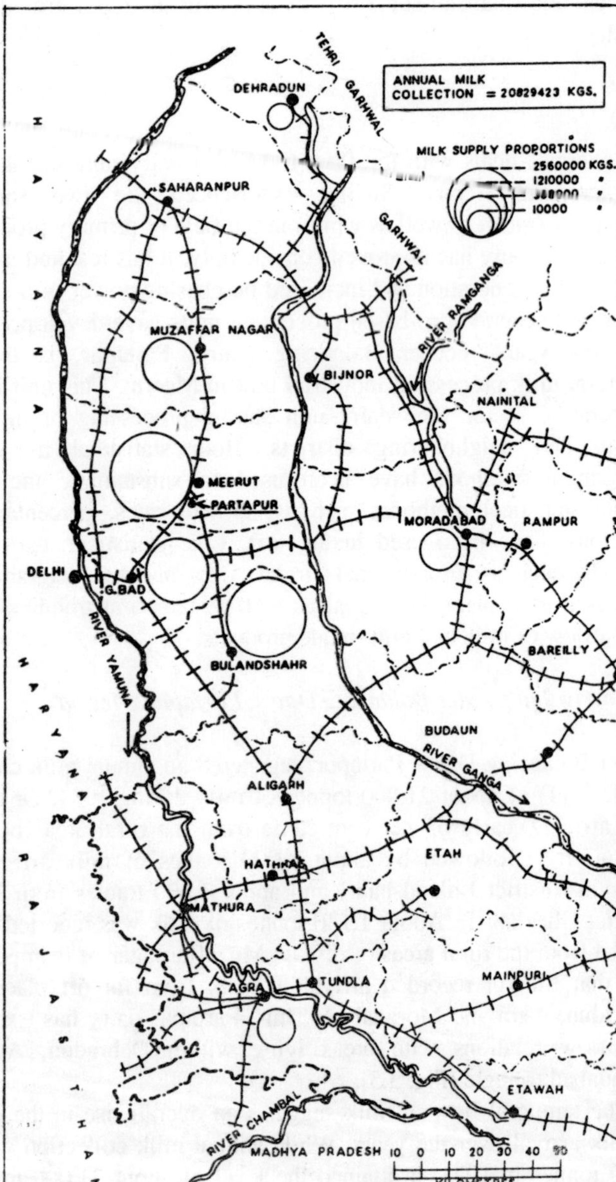

FIG. 3.5: Inflow of milk in Feeder Balancing Dairy, Partapur, Meerut — 1984 (special trends).

arrival was recorded during 1976-77 and 1980-83. It is, thus, clear that with the two trends of sharp rise, there was one trend of nominal decline in milk arrival during the period under study. These (up and down) trends are due to several factors, like, drought conditions, diseases, inadequate vetenary services, poor links, etc.

TABLE 3.11: Meerut City — Arrival of Milk at Feeder Balancing Dairy, Partapur

	Trends of arrival (in tonnes)						
	Annual (1973-84)	*Monthly (1984)*		*Spatial (1984)*			
Year	*Quantity*	*Months*	*Quantity*	*Places*	*Quantity*	*Percent-age*	*Rank*
1973	2,830	January	1,803	Meerut	13,008	62.45	1
1974	3,109	February	1,660	Muzaffarnagar	1,133	5.44	4
1975	4,605	March	1,754	Gulaothi	3,960	19.01	2
1976	7,200	April	1,319	Masoori (G.Bad)	1,989	9.55	3
1977	17,430	May	959	Saharanpur	80	0.39	7
1978	14,136	June	892	Moradabad	365	1.75	5
1979	23,842	July	994	Agra	291	1.40	6
1980	26,373	August	1,012	Dehradun	3	0.01	8
1981	24,137	September	1,737				
1982	18,455	October	2,190				
1983	17,601	November	3,038				
1984	20,829	December	3,471				

Monthly trend (Fig. 3.6) of milk arrival explains the effect of seasonality in the milk arrival. The milk arrival ranged between about 900 (892) tonnes in June to about 3,000 (3,400) tonnes in December, yielding the range of a variation of about 2,4000 tonnes of milk. Arrival of milk declines during January to June whereas it increases during the period July to December.

Outflow of Dairy Products

Outflow of milk and milk-made items explains the functional (milk) relationship of Meerut with other places. It also explains the extent of dependency on Meerut for milk and milk-made items. 1984 registered (Table 3.12) about 11,000 tonnes of milk, 469 tonnes of butter, 362 tonnes of *ghee* and about 130 tonnes of skimmed milk powder as outflow from Meerut. While butter and *ghee* were supplied to Delhi and to six places in Uttar Pradesh, skimmed milk powder was supplied to seven places in Uttar Pradesh alone. Outward supply of milk products was scattered over the region. In terms of outflow of milk, Lucknow ranked second and accounted for a major share. Distribution pattern seems to be affected by the demand of the product. Outflow of milk was recorded in seven places. Of the total outflow of milk, nearly 30 per cent was supplied to the city of Lucknow followed by 28.43 per cent to Meerut city. Delhi ranked third and recorded about 21 per cent of the outflow of milk. While Ghaziabad registered about 11 per cent of the total supply, Kanpur registered about eight per cent. Dehradun and Varanasi centres have also registered nominal receipt of milk from Meerut during the period under study. It is, thus, apparent that weight does not tend to decrease with distance as Lucknow recorded maximum receipt of milk from Meerut. Demand is, therefore, most important factor explaining the extent of outflow of milk (Fig. 3.7)

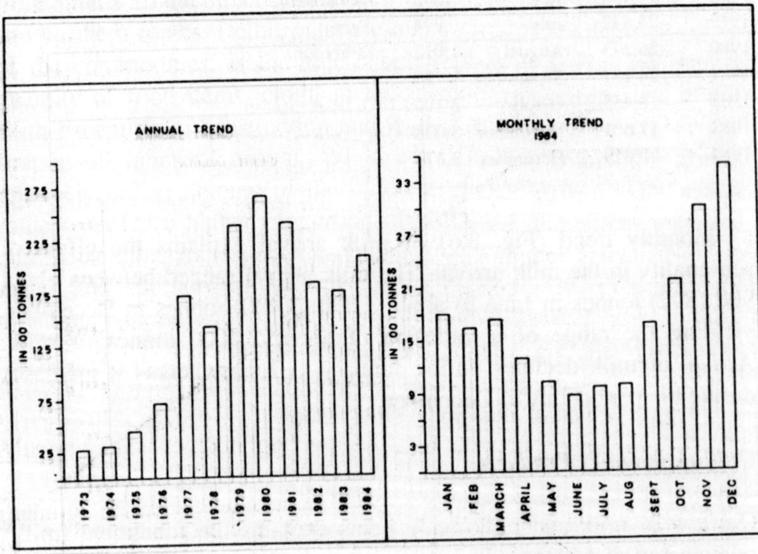

FIG. 3.6: Meerut city — Inflow of milk at Feeder Balancing Dairy, Partapur, Meerut (temporal trends).

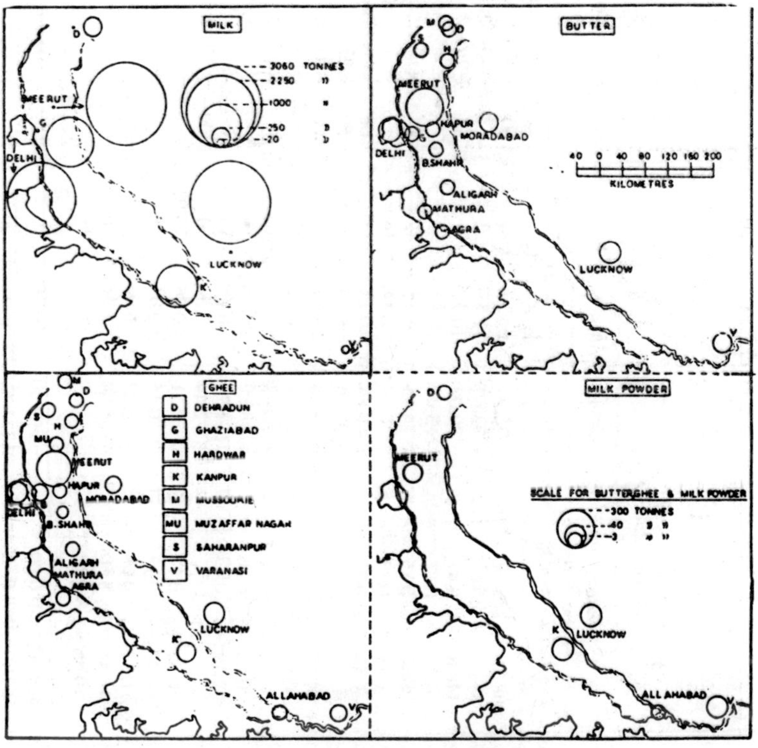

FIG. 3.7: F.B.D., Partapur, Meerut — Outflow of dairy products — 1984.

Butter is supplied to about 15 places from Mussoorie to Varanasi. Most of the butter supply, about 64 (63.57) per cent is for the Meerut city itself, followed by Delhi about 16 (15.61) per cent. Lucknow, Varanasi and Moradabad registered nearly 18 per cent of the total butter supply. Remaining about three per cent of supply is directed towards nine centres.

Similar to that of butter, *ghee* supply was also recorded for 17 centres along Mussoorie-Varanasi axis. Of the total *ghee* outflow, about 61 per cent was consumed in the Meerut city. Next in order of *ghee* supply were Delhi and Lucknow which accounted for over 22 per cent of the total outflow. Kanpur, Varanasi, Bulandshahr and

TABLE 3.12: Meerut City – Outflow of Milk and Milk-Made Items — 1984

(Weight : In Tonnes)

Places	Milk Arrival	Milk Percentage	Milk Rank	Butter Arrival	Butter Percentage	Butter Rank	Ghee Arrival	Ghee Percentage	Ghee Rank	Skimmed milk powder Arrival	Skimmed milk powder Percentage	Skimmed milk powder Rank
Meerut	3,044.00	23.43	2	298.15	63.57	1	222.00	61.32	1	24.00	18.60	4
Delhi	2,283.00	21.32	3	73.23	15.61	2	43.00	11.88	2	—	—	—
Lucknow	3,223.00	30.10	1	44.16	9.42	3	39.00	10.77	3	38.50	29.84	1
Ghaziabad	1,165.00	10.88	4	0.35	0.07	15	3.00	0.83	9	—	—	—
Saharanpur	—	—	—	4.00	0.85	6	4.15	1.15	7	—	—	—
Roorkee	—	—	—	1.53	0.33	8	0.49	0.14	14	—	—	—
Dehradun	162.00	1.51	6	1.51	0.32	9	0.83	0.23	13	1.00	0.78	6
Kanpur	808.00	7.55	5	—	—	—	21.00	5.80	4	32.50	25.19	2
Varanasi	23.00	0.21	7	24.35	5.19	4	14.00	3.87	5	30.00	23.26	3
Mussoorie	—	—	—	0.80	0.17	10	1.14	0.31	12	—	—	—
Muzaffarnagar	—	—	—	—	—	—	0.47	0.13	15	—	—	—
Agra	—	—	—	0.28	0.06	—	3.10	0.86	8	—	—	—
Aligarh	—	—	—	2.19	0.47	7	1.74	0.48	11	—	—	—
Allahabad	—	—	—	—	—	—	2.49	0.69	10	3.00	2.33	5
Hapur	—	—	—	0.52	0.11	13	0.21	0.06	16.5	—	—	—
Mathura	—	—	—	0.60	0.13	11	0.15	0.04	18	—	—	—
Bulandshahr	—	—	—	0.37	0.08	14	0.23	0.06	16.5	—	—	—
Moradabad	—	—	—	16.41	3.50	5	5.00	1.38	6	—	—	—
Hardwar	—	—	—	0.55	0.12	12	—	—	—	—	—	—
Total	10,708.00	100.00		469.00	100.00		362.00	100.00		129.00	100.00	

Saharanpur were in the third order of *ghee* supply. The proportion of *ghee* receipt in these places ranged between one (1.15) per cent in Saharanpur to about six (5.80) per cent in Kanpur. Remaining centres registered less than one per cent of the total supply.

Distribution of skimmed milk powder (s.m.p.) was limited to a few places. Of the total outflow of s.m.p., nearly 30 per cent was directed to Lucknow, 25 per cent to Kanpur, 23 per cent to Varanasi, 19 per cent to Meerut city and remaining about four per cent to Allahabad and Dehradun.

The above facts reveal that in case of dairy industry, milk was the main item of supply and was largely related to the demands of the metropolitan centres like Delhi, Lucknow and Kanpur. Growing urban populations of these centres have increased the demand for milk and other dairy products. Milk products recorded limited supply possibly because of the low demand. Thus, while milk has become the essential household demand, milk products have to find place in consumer demands. Limited demand for milk-made items is partly due to its higher rates and partly due to low purchasing power of the majority of the consumers.

Conclusion

The foregoing facts reveal that industrial activity is gaining gradual significance in the economic sphere of Meerut city. One of the favourable factors for industrial growth is the easy access and availability of raw material. Since raw material linkages are weak, industrial production varies significantly. While industrial products recorded regular supply with other parts of the country, it recorded seasonal variations in the arrival of products. The products are sold in the national market at a fairly high rates whereas products purchased from household manufacturers are at a considerably low rates. In the absence of industrial complexes, producers get only marginal benefit of their products. As a result, household producers/manufacturers are left with no choice than to go for the distress sale of their products. Apart from price, surplus production also awaits storage and assured purchase of their products. Due to missing links (forward and backward) in the organisational set-up of industries, production fails to co-ordinate the hierarchic chain of settlements. Industrial units mostly run short of the raw material. On the contrary, surplus production awaits early supply. Prevailing dichotomy of industrial development has led to the identification of distorted space relations of Meerut city. Since most of the industrial units are small, unregistered and unorganised functional gaps and income inequalities tend to grow. Some of the industrial products like sports goods and textiles have established their credit in the national as well as international markets. However, planned attempts are needed to increase the export level of these industrial products. Raw material banks, industrial corporations, liberal financial support and modernisation of industrial technology would transform the industrial structure of Meerut city.

NOTES

1. *Directory of Industries*, Meerut, 1984.
2. Coefficient of variation $= \frac{\sigma}{\bar{x}} \times 100$
3. Coefficient of correlation $= \frac{\sum XY}{(\sigma x)(sigmaay)}$ and]
4. Regression equation = *a* + *bx*
5. Supply details are not available for all places in Uttar Pradesh. As such, available details of supply to Kanpur, the headquarters of U.P. Handloom Intensive Development Project have been analysed to give an idea of the importance of the State.
6. The details are based upon the pamphlet issued by District Industry Centre, Meerut, on 30th April, 1984.

4

SPATIAL AND TEMPORAL MOVEMENT OF POPULATION

INTRODUCTION

THIS CHAPTER deals with human interactions which involve the movement of people from one place to another. Interaction potentiality of a place is expressed in terms of movement generated for employment, education, services, marketing, entertainment and other such functions. To recognise the intensity of interaction and extent of human dependence, it is essential to analyse the patterns of human movement. Distance is the most important factor explaining extent of human movements. It is a limiting element of human movement over places and regions. However, with the introduction of modern means of transport and communication, physical distance is reduced considerably. Broad themes, under which human interactions have been analysed, are the spatial and temporal dimensions of human movements, purpose of movement, frequency of movement and modes of transport used in human movements. Meerut city, besides being a regional node for western Uttar Pradesh, also acts as an important feeder to the National Capital – Delhi. Thus, flow of surplus produce, metropolitan demand and growing urban and regional economy have jointly accelerated the pace of human interactions in case of Meerut. Highways and railways are the main links through which human movements are made.

Data used in this exercise refer to the period 1975-84 in case of railways and to the period 1981-84 in case of highways. While time series data were used for analysing temporal trends, cross-sectional data (1984) were used for analysing trends of monthly movements, earnings and other related factors. Intensive field survey was conducted in the months of May, August and November 1986. A week-long survey was launched separately at railway stations, bus terminals and all entry points to the city. Survey included questions related with the place of origin and destination, purpose of movement, frequency of movement, mode of movement and problems of movements. Data, thus, collected were tabulated and processed through cartographic and statistical methods. Cartographic methods used here are flow diagrams, bar and line graphs and proportional circles whereas statistical methods used in the exercise are intensity of interaction, distance decay, deviations, correlations and regression coefficients. Broad themes of the chapter are discussed below separately :

Place of Movement

During 1984, Meerut city recorded human interaction with 203 places spread over a distance range of 1,600 km. Of the total places with which human movements are recorded, nearly 28 per cent fall within a range of 100 km, 23 per cent within 101-200 km, 20 per cent within 201-300 km, 14 per cent within 301-400 km and six per cent within 401-500 km. Remaining about 10 per cent of the places lie within a distance range of 501-1,600 km. Nearly 70 per cent of the places, thus, fall within a distance of 300 km (Table 4.1).

Amount of human movement to these places varies significantly. About 19.56 million people have registered movement (to and from) with Meerut city during 1984. As much as 71 per cent of the total human interaction was limited to a distance of 100 km, 16 per cent to 101-200 km and nine per cent to 201-300 km. Thus, over 96 per cent of the human interaction was confined to a range of 300 km. Amount of interaction declines sharply beyond this range, the share of interaction being about four (3.82) per cent.

Marked variations have also been recorded in the human movements through highways and railways. Of the total place interaction, links through highways and railways were about 51 and 49 per cent whereas of the human interaction share of movement were 88 and 12 per cent, respectively. Nature of place interaction through

TABLE 4.1: Meerut City — Places and Amount of Human Interaction — 1984

Distance from Meerut city (in km)	Place Interaction		Share of place interaction through		Amount of human movement	Percentage	Share of human movement through	
	Number	Percentage	Rail	Road			Rail	Road
1	2	3	4	5	6	7	8	9
up to 50	40	19.70	50.00	50.00	97,26,597	49.72	7.63	92.37
51-100	17	8.37	88.24	11.76	41,86,236	21.40	21.65	78.35
101-150	24	11.82	33.33	66.67	20,08,294	10.27	8.45	91.55
151-200	22	10.84	22.73	77.27	10,54,647	5.39	5.58	94.42
201-250	23	11.33	26.09	73.91	17,26,436	8.82	0.81	99.19
251-300	15	7.39	40.00	60.00	1,13,471	0.58	21.41	78.59
301-350	14	6.90	46.15	53.85	2,51,816	1.29	18.97	81.03
351-400	15	7.39	33.33	66.67	48,697	0.25	64.60	35.40
401-450	7	3.45	71.43	28.57	1,67,474	0.86	82.68	17.82
451-500	6	2.96	50.00	50.00	1,43,477	0.73	59.39	40.61
501 and above	20	9.85	95.24	4.76	1,34,524	0.69	100.00	—
Total	203	100.00			195,61,669		12.03	87.97

Source: Computed from the records of Railways and Roadways. For details refer to Aappendix-2A and 2B.

railways presented a bimodal curve in which medium distance (25.25%) had relatively lesser proportion compared to the short and long distances (39.39%). Of the total human movement through railways, nearly 70 per cent was for short distance (up to 100 kms.), about 19 per cent for long distance (over 300 km) and remaining 11 per cent for medium distance. On the other hand, highways presented a normal curve for place interaction which is characterised by 57 per cent of places within medium distance slab, 22 per cent within long and remaining 21 per cent within short distances. On the other hand, amount of human movement through highways presented a declining trend characterised by 71 per cent interaction for short distance, 27 per cent for medium and remaining two per cent for long distances. Results reveal that both the ways of movement are almost equally important (99 places through railways and 104 places through highways) in case of place interaction. But there are variations in the number of places, approached through these modes, over different distance slabs. Through railways, share of place interaction remains high over short and long distances but through highways, it remains high only in the medium distance slab. Unlike place interaction, amount of human movement presented the lopsided distribution. While over short and medium distances amount of human movement remains relatively higher through highways, over long distances it remains higher through railways.

About 20 (19.65) million people were registered on the move during 1984. As a result of this massive human interaction, Meerut city recorded an earning worth Rs. 106 (106.1) million. Of the total human interaction, share of movement through railways and highways was 12 (12.03) and 88 (87.97) per cent, respectively. Whereas corresponding share in the earning was 30 (29.86) and 70 (70.13) per cent. These facts explain that despite low share of human movement, railway contributes a sizable sum in the annual earnings. As discussed earlier, most of the long range traffic moves through railways. Consequently per ticket average earning becomes more through railways compared to the highways.

Proportion of human movement does not remain uniform throughout the year. Amount of movement increases significantly during summer and declines proportionately during rainy and winter seasons (August, November and December). Thus, there are seasonal variation in the human movement. Temporal trends of human movement reveal that Meerut city recorded a decline in the traffic by

17.76 per cent during 1981-82 and subsequent rise by five (5.45) and 13 (13.09) per cent during 1982-83 and 1983-84. Similarly, trends of earning, through human movement, registered a continuous rise from 1981 to 1984. The annual earnings rose by about 10 (9.91) per cent during 1981-82, 21 (20.96) per cent in 1982-83 and 25 (24.74) per cent during 1983-84. It is, thus, clear that place interaction has more regular than random distribution, whereas human interaction has more clustered than regular distribution. The two distributions explain that large number of people moves over short distances and visit only a few places and small number of people moves over long distances and visit many places. Interaction is one of the important sources of earning for Meerut city. It is, therefore, necessary to analyse the details of railways and highways in terms of geographical coverage, relative importance of the routes, temporal trends of human movement and earnings.

Railways

Spatial pattern of interaction through railways reveals (Fig. 4.1) that Meerut city had a coverage up to Saharsa in the east, Nagpur in the south-east, Bombay in the south, Ajmer in the south-west, Atari in the north-west and Jammu Tawi in the north. Based on the amount of human interaction, entire coverage may be sub-divided into three sets. First set has high intensity of human interaction and is nearest to the Meerut city. Muzaffarnagar, Hapur and Delhi are the coordinates of the first set and the places lying within this triangle register maximum interaction with Meerut. The second set begins after the first set and includes areas having moderate intensity of interaction. This set covers areas up to Lucknow, Allahabad, Bombay and Amritsar. The pressure of interaction on these routes remains lesser compared to the pressure experienced in the first set. The third set includes those centres which register low intensity of human interaction. Centres lying along Lucknow-Saharsa, Allahabad-Mugal Sarai, Agra-Nagpur, Tundla-Farrukhabad, Delhi-Ajmer and Saharanpur-Dehradun sections fall in this set. Human interaction along these sections is less than 10,000 persons per annum.

Amount of monthly movement explains large variations. While minimum movement was recorded in the month of November, i.e., 14 lakh persons, maximum was recorded in May, i.e., about 24 lakh passengers. Relatively lesser amount of human movement was recorded during November to February whereas relatively greater amount of interaction was recorded during April and May. Remaining months of the year recorded an average amount of human interaction. Similar to that of monthly trend annual trend of interaction presents large variations. Meerut city recorded the movement of about 3 (3.24) million passengers during 1975. There was a regular increase in the number of passengers until 1979. By this period, amount of interaction reached the level of more than 4 (4.44) million passengers. The rate of average increase for the period (1975-79) was 9.3 per cent per annum. The period between 1979 to 1984 witnessed an overall decline in the amount of movement, i.e., from 4.44 to 2.44 million persons. But the periods 1980-81 and 1983-84 registered an increase of 20.00 and 7.49 per cent, respectively. On the contrary, a decline of 10 (9.91), 39 (38.75) and 23 (22.79) per cent were recorded for the respective periods of 1979-80, 1981-82 and 1982-83.

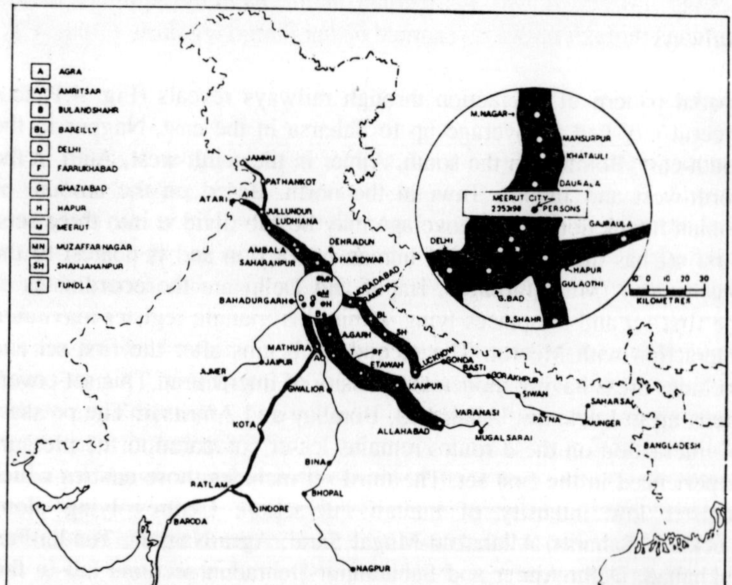

FIG. 4.1: Meerut city — Passengers movement through railways — 1984.

Amount of human movement influences the extent of earnings significantly. Railway earnings ranged between about Rs. 2 (17.90) million in November to about 3 (3.46) million in May through ticket bookings. While April, May and June recorded high earnings, November, January and February recorded low earnings. Remaining months of the year registered close to the average earning, i.e., Rs. 2.64 million. Despite declining trend of passengers for the past few years, trends of annual earning registered all through (1975-84) a rising trend. Railway earnings, during 1975, were about Rs. 11 (10.96) million which rose to the level of about Rs. 32 (31.67) million by the turn of 1984. All through this period, there has been a positive growth in the earnings except for the period 1978-79 which recorded a negative growth of 3.45 per cent. The entire period between 1975-84 registered a general growth rate of 13.05 per cent per annum. It is, thus, clear that despite ups and downs (variations) in the monthly earnings, annual earnings through railways recorded upward trend regularly (Table 4.2).

Highways

Interaction through highways recorded coverage up to Mathura, Agra, Etawah, Farrukhabad and Lucknow in the east, Chandigarh, Dehradun, Rishikesh, Kotdwar, Haldwani and Tanakpur in the north, Paliya and Lakhimpur in the east, Kurukshetra and Kerala in the west and Delhi in the south. The high intensity of interaction was recorded along Hardwar-Delhi axis. Moderate intensity of interaction was recorded along Meerut-Shahjahanpur and Meerut-Agra routes. Low intensity of interaction was recorded along Aligarh-Farrukhabad, Agra-Etawah, Hathras-Mathura, Shahjahanpur-Lucknow, Shahjahanpur-Paliya, Rampur-Tanakpur, Bijnor-Kotdwar and Saharanpur-Chandigarh routes. It is, thus, clear that human interaction through highways remains quite significant in all the directions except towards west. Limited place interaction towards the west is due to the limited links of the western boundary of the district. The fact that there are a few bridges on the river Yamuna, the western limit of the district, amount of interaction towards west remains low and for a limited distance. Hardwar-Delhi axis is, thus, the main artery of interaction and accounted for more than half of the total interaction. It is established that the extent of interaction is proportional to the connectivity of a place and the interaction is also influenced by the order of a place and the order of a route (Fig. 4.2).

TABLE 4.2: Meerut City — Temporal Trends of Railway Bookings and Earnings

Months	Monthly trends (1984)				Year	Annual trends (1975-84)			
	Passenger tickets (in millions)	*Percentage to total tickets issued*	*Earning (in million rupees)*	*Percentage to total earning*		*Passenger tickets (in millions)*	*Growth rate*	*Earning (in million rupees)*	*Growth rate*
January	0.18	7.38	2.31	7.29	1975	3.24	—	10.96	—
February	0.17	6.97	2.21	6.98	1976	4.10	26.54	14.90	35.95
March	0.21	8.61	2.68	8.46	1977	4.33	5.61	15.13	1.54
April	0.23	9.43	2.90	9.16	1978	4.34	0.23	15.64	3.37
May	0.24	9.84	3.46	10.93	1979	4.44	2.30	15.10	—3.45
June	0.21	8.61	3.28	10.36	1980	4.00	—9.91	16.74	10.86
July	0.22	9.01	2.67	8.43	1981	4.80	20.00	20.25	20.97
August	0.22	9.01	2.56	8.08	1982	2.94	—38.75	23.64	16.74
September	0.22	9.01	2.68	8.46	1983	2.27	—22.79	27.76	17.43
October	0.22	9.01	2.80	8.84	1984	2.44	7.49	31.67	14.09
November	0.14	5.74	1.79	5.65					
December	0.18	7.38	2.33	7.36					
Total	2.44	100.00	31.67	100.00					

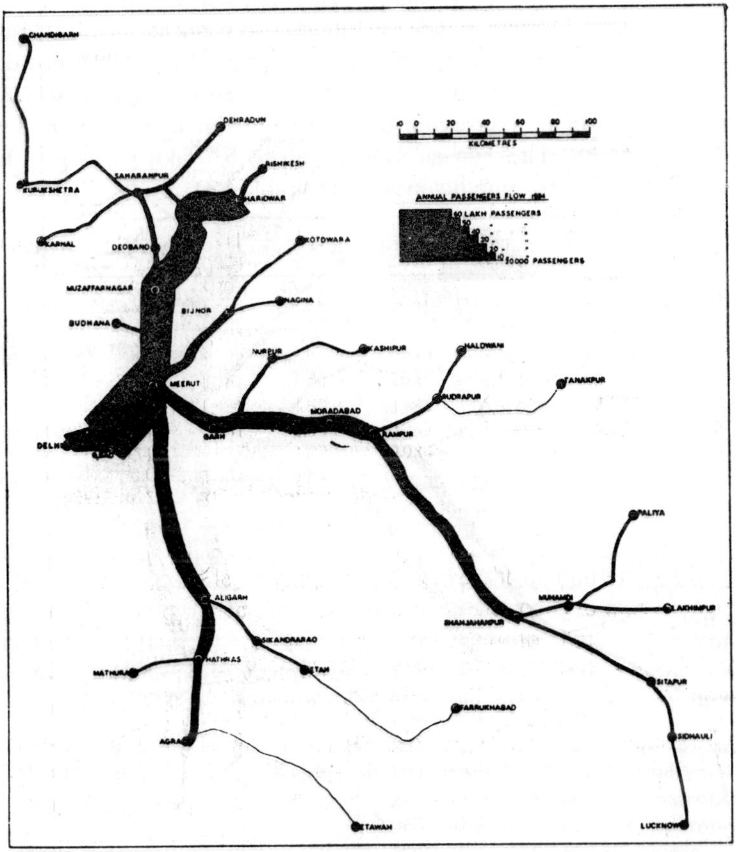

FIG. 4.2: Meerut city — Passengers movement through highways — 1984.

Temporal trends reveal marked monthly variations in the human movements through highways. While August, November and December accounted for relatively lesser proportion of human movement, March, May, June and July accounted for relatively greater proportion of human movement. Remaining five months registered an average amount of movement. The amount of human movement ranged between 1.25 million persons (in August) to 1.60 million persons (in May). The average monthly movement was calculated as 1.43 million persons. Apparently, share of human movement rises during summer, falls down during rainy and remains average during winter season. Unlike trends of monthly movement, annual trends registered the regular rise for the past four years. While 1981 recorded only about 15 (15.26) million passengers, 1984 recorded 17 (17.21) million passengers registering an annual growth rate of about four (4.25) per cent. Thus, on an average, about 6.5 lakh passengers are added to the human mobility every year (Table 4.3).

TABLE 4.3: Meerut City — Temporal Trends of Highway Bookings and Earnings

Months	*Monthly trends (1984)*				*Annual trends (1981-84)*				
	Passenger tickets (in millions)	*Percentage to total tickets issued*	*Earning (in million rupees)*	*Percentage to total earning*		*1981*	*1982*	*1983*	*1984*
January	1.42	8.25	5.81	7.81	Tickets issued	15.26	13.55	15.12	17.21
Februaary	1.49	8.66	6.14	8.25	(in million)				
March	1.55	9.01	6.50	8.74	Growth	—	–1.71	+1.57	+2.09
April	1.38	8.02	5.96	8.01	Growth rate	—	–11.21	+11.59	+13.82
May	1.60	9.30	7.21	9.69	Income through	43.71	46.66	57.28	74.41
June	1.52	8.83	6.62	8.90	human movement				
July	1.54	8.95	6.73	9.04	(in million Rs.)				
August	1.25	7.26	5.27	7.08	Growth	—	2.95	10.62	17.13
September	1.41	8.19	5.97	8.02	Grwoth rate	—	6.75	22.76	29.91
October	1.44	8.37	6.61	8.88					
November	1.25	7.26	5.58	7.50					
December	1.36	7.90	6.01	8.08					
Total	17.21	100.00	74.41	100.00					

Highways registered high earnings through human movements. The monthly earnings ranged between over Rs. 5 (5.27) million in August to over Rs. 7.21 million in May. The average monthly earning during 1984 was Rs. 6.20 million. Trends of annual earnings registered a growth rate of 22 per cent. While 1981 recorded the earnings of about Rs. 44 (43.71) million, 1984 recorded the earnings of Rs. 74 (74.41) million. Earnings, thus, rose at an average rate of about Rs. 10 (10.23) million per year. It is, therefore, established that despite marked variations in the amount of monthly movement and earnings. There was a regular rise in the amount of human movement and earnings (Fig. 4.3).

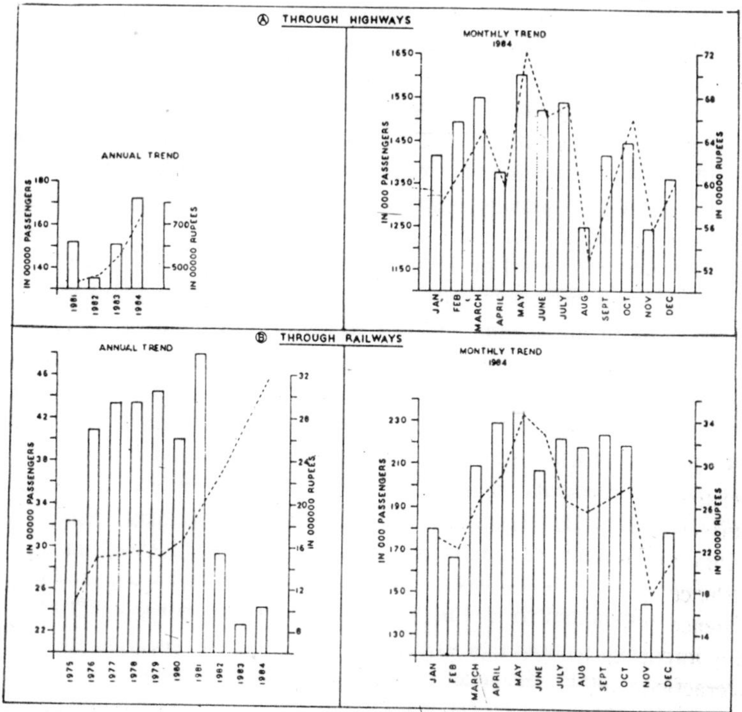

FIG. 4.3: Meerut city — Passengers movement and earnings through highways and railways.

Distance and Purpose of Movement

As discussed above, it is now clear that the amount of human interaction varies from place to place depending upon the distance and the order of a place. However, it remains to be investigated as to what for such movements are made? It is known that interactions are purpose oriented. The purpose of interaction changes over time and place. This section deals with the purpose of movement over different distance slabs. Distance is the most important geographical reality explaining human movement. For general purposes, short distance and for special purposes, long distance interactions are made. To have a comprehensive idea about the purpose of movement, a detailed field survey was conducted at the railway stations (Meerut city and Cantt.), bus and taxi stands. Field survey revealed that of the total interaction, through railways, nearly 70.07 per cent was for short distance, about 14.71 per cent for medium distance and remaining over 15.22 per cent for long distance. Further, maximum amount of movement was for social purposes (48.17%) including movements for attending marriages and festival celebrations and for meeting the relations. Thus, large amount of human movement is for maintaining social relations. Besides being unproductive economically, it is also a huge burden on the income of the society. Next to social interaction, is the interaction for work places (service interaction). This type of interaction accounted for one-third of the total. Interaction for the purposes, like, trade and commerce, accounted for about 14 per cent and ranked third in the order of interaction. Nearly five per cent of the total interaction was for legal advice, political, administrative and religious purposes. The overall trend of interaction does not remain the same at all levels of slab interaction. For instance, short distance movement registered maximum interaction for work place (45%) whereas medium and long distance movements registered maximum interactions for social purposes. On the contrary, short distance movement registered minimum amount of interaction for legal advice (2.25%), medium distance movement for work place interaction (9.91%) whereas long distance movement for political and administrative purposes (2.86%). It is important to note that as compared to short and medium distance interactions, long distance interaction registered relatively higher proportion of interaction for legal advice, political and administrative purposes. It is, thus, clear that intensity of interaction increases with distance for the purposes, like, social, political, administrative and legal advice. On the contrary intensity of interaction decreases with distance for the purposes, like, workplace mobility, business and trade interactions (Table 4.4).

Movement through highways, on the other hand, reveals that nearly 53 per cent of the total interaction was directed towards short distance, 30 per cent to medium distance and remaining about 17 per cent to long distance. Purposes, for which these distances were covered, accounted for 30 per cent social interaction, 25 per cent trade, 24 per cent legal advice and remaining nearly 20 per cent for entertainment, medical, alcoholic drinks, education and for search of jobs. Interaction through highways is for a variety of purposes over limited distance (up to 600 km). Compared to the interaction through railways which is for a few selected purposes over long distances (upto 1,600 km) (Table 4.5).

TABLE 4.4: Meerut City — Purpose of Movement Through Railways

	Distance slab (in km)			
	As percentage to total movement			
Purposes	*Short distance up to 100 km*	*Medium distance 101-400 km*	*Long distance 401-1600 km*	*Percentage to total*
Legal advice	2.25	—	4.77	2.30
Service	45.50	9.91	—	33.34
Business	12.25	23.58	11.67	13.83
Social	37.25	66.51	80.70	48.17
Religious	2.75	—	—	1.93
Political & Administrative	—	—	2.86	0.43
Total percentage	100.00	100.00	100.00	100.00
Total (Actual movement)	16,48,790	3,46,207	3,58,201	23,53,198
Percentage	70.07	14.71	15.22	100.00

Source: Based on the field enquiry conducted at Meerut railway station.

Slabwise interaction for various purposes revealed interesting results. While legal advice ranked first in short distance interaction, social and trade figured on the top in medium and long distance interactions. It is further established (through Table 4.5) that the share of trade to the total slab interaction increased corresponding to the distance whereas the share of social interaction decreased corresponding to the distance. Legal advice ranked first in short distance interactions. It is followed by the interactions for trade, social and work-place movements. Interaction for entertainment and medical purposes accounted for seven per cent each whereas interactions for alcoholic drinks and education together accounted for about six (6.35) per cent. Medium distance slab was characterised by the maximum of social interaction (nearly 60%) followed by movement for trade and religious purposes (37.67%). Work-place movement and movement for education purposes registered the minimum scores, i.e., nearly one per cent each. It is important to note that interactions for legal advice, entertainment, medical and alcoholic drinks did not register any amount of interaction in this slab. Due to fairly widespread distribution of these functions/facilities, interaction remains limited to short distance only. Long distance interaction becomes much more selective. Here maximum interaction was recorded for trade followed by interaction for social purposes. Second, in order of importance, was the movement for the search of jobs. This type of interaction accounted for nine per cent of the total slab interaction. Such movements often turn to be the unserved purpose of interactions. People from Bihar, parts of of eastern Uttar Pradesh and parts of hill U.P. interact with Meerut for search of jobs. Sometimes they have to move to other places, if the purpose remains unserved here. Third interactions in order of importance are for religious, political and administrative purposes. These purposes accounted for about one (1.30) per cent of the total interaction.

TABLE 4.5: Meerut City — Purpose of Movement Through Highways

Purposes	*Distance slab (in km)* *As percentage to total movement*			*Percentage to total*
	Short distance (up to 100 km)	*Medium distance (101-200 km)*	*Long distance (201-600 km)*	
Legal advice	45.17	—	—	23.86
Entertainment	7.30	—	—	3.86
Medical	7.30	—	—	3.86
Business	14.05	28.89	50.87	24.72
Alcoholic drinks	3.65	—	—	1.93
Education	2.70	1.22	—	1.79
Social	10.79	59.89	38.70	30.41
Service (Workplace)	9.04	1.22	—	5.15
Religious	—	8.78	0.87	2.82
Search for job	—	—	9.13	1.53
Political and administrative	—	—	0.43	0.07
Total percentage	100.00	100.00	100.00	100.00
Total (Actual movement)	90,90,814	52,40,898	28,76,756	1,72,08,471
Percentage	52.83	30.45	16.72	100.00

Source: Based on the field enquiry conducted at Meerut Depot.

Thus, amount and purpose of interaction presented large variations through these means (Railways + Highways). Further, there are significant variations within each means of movement. Purpose of interaction varies with distance slab. For instance, amount of movement decreases with distance and the purpose of movement becomes selective and of higher order with the distance. Movement of population is maximum for maintaining social relations and its proportion increases towards long distances. Besides common generalization, there are a few specific generalizations as well. As compared to the railways, interaction through highways extended to a shorter distance satisfying multi-purpose requirements. While most of the work-place interaction was through railways, most of the interaction for legal advice was through highways. Similarly interaction for the purposes, like, social, political and administrative, was more through railways whereas interaction for trade, alcoholic drinks and entertainment was relatively more through highways. Due to frequent availability of the alternate modes of highway transport (Private as well as Government buses along with other passengers carriers), specialisations have developed in the various fields of interactions. Railway transport, on the other hand, are limited and are available over the fixed timings. As a result of the existing transport system, short and medium distance interactions are made mostly through highways and are for a variety of purposes whereas long distance interactions are made mostly through railways and are for selected higher order purposes. Despite complimentarity of the means of transport, gaps remain to be fulfilled for generating all round balanced interactions. Frequency of the train services need to be increased to cope with the growing pressure of human movement. Similarly western part of the district requires adequate improvement in the highway connectivity (Fig. 4.4).

Frequency of Movement

Frequency of movement through railways and highways explain the intensity of interaction with Meerut city. Amount of human movement has been analysed under the daily, monthly, seasonal, half-yearly, yearly and rare frequencies of the human movements. Field survey carried out at all terminals of Meerut city revealed that nearly 57 (57.01) per cent of the movement to the city was daily frequency. Of the daily movement of the city, over 85 per cent was through highways and remaining through railways. It was estimated that about 30,000 people made daily movement to the city during 1984 (Table 4.6). Out of annual traffic of 19.56 million passengers about 11.15 million passengers were daily commuters. Twenty five (25.63) per cent of the total interaction was of monthly frequency, the proportionate share of interaction through highways and railways being 88 (87.97) and 12 (12.03) per cent. It was estimated that over 4 (4.18) lakhs passengers recorded monthly interaction with the city during 1984. Seasonal and half-yearly movements accounted for 10 (9.48) and five (4.56) per cent. The proportionate share of interaction through highways and railways remained the same, i.e., 88 and 12 per cent for monthly, seasonal and half-yearly frequencies. About three (2.79) per cent of the total interaction registered annual frequency of movement with the ratio of 93 (93.20) and seven (6.80) per cent through highways and railways. About 5.46 lakh persons were estimated to have made this type of movement during 1984. Frequency of rare movements was less than one (0.53) per cent of the total interaction and were conducted mostly through highways (94.81%). About one lakh (1.03) lakh passengers were estimated to have made this type of movement during 1984. Analysis further revealed that highways played relatively more important role in maintaining the frequency of movements to and from the city. Amount of interaction is proportional to the frequency of

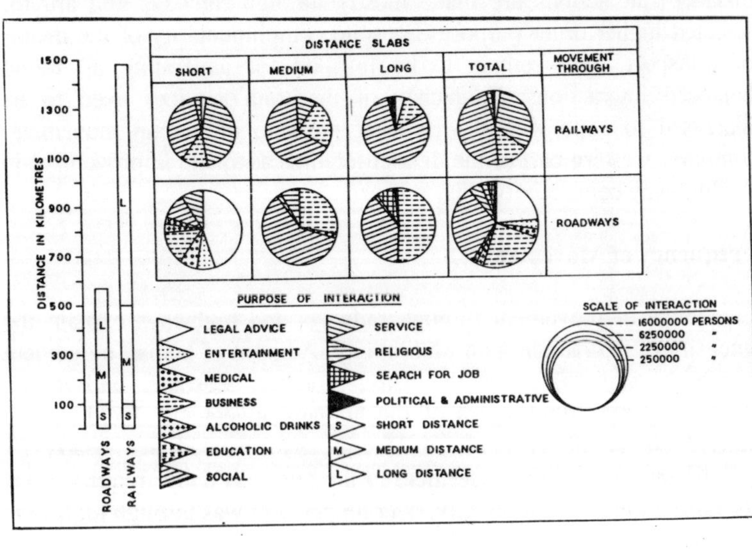

FIG. 4.4: Meerut city – Distance and purpose of movement.

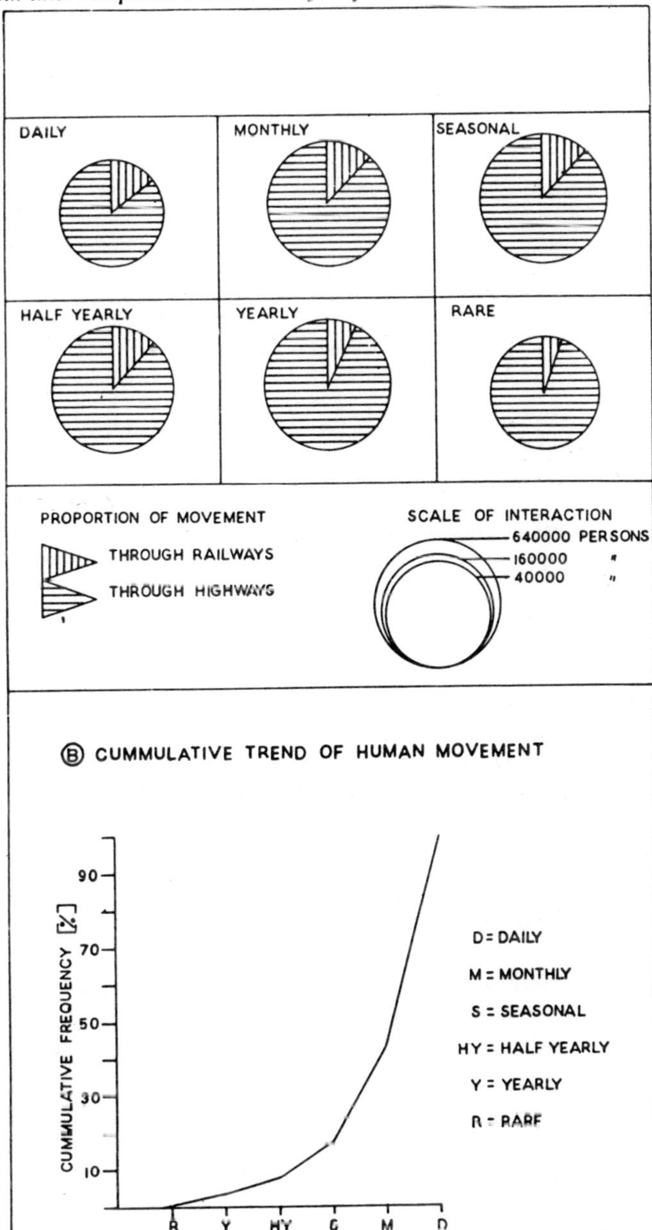

FIG. 4.5: Meerut city — Amount of human movement and cumulative trend of human movement.

TABLE 4.6: Meerut City — Frequency of Human Interaction — 1984

Frequency group	*Amount of interaction*	*Cummulative amount of interaction for the calendar year*	*As percentage to the annual total*	*Proportion of interaction through*	
				Railways	*Highways*
Daily	30,552	1,11,51,638	57.01	14.60	85.40
Monthly	4,17,846	50,14,160	25.63	12.03	87.97
Seasonal	6,17,934	18,53,802	9.48	12.03	87.97
Half-yearly	4,46,348	8,92,696	4.56	12.03	87.97
Yearly	5,46,028	5,46,028	2.79	6.80	93.20
Rare	1,03,345	1,03,345	0.53	5.19	94.81
Total		1,95,61,669	100.00	12.03	87.97

Note: These figures are based on the records of passenger movement through highways and railways for the year 1984. However, frequency information is based on the sample survey conducted during 1986. Projection is based on the assumption that proportion of movement would be the same in 1986 as it was during 1984.

movement. Similarly, ratio of the interaction decreases with the frequency of movement in case of highways. On the contrary, it increases in case of railways. It is important to note that half-yearly and yearly frequencies were not as important as seasonal frequency because it ranked third after daily and monthly frequencies. Frequency distribuiion of movement presents a declining trend where frequency of daily movement formed the peak of interaction (Fig. 4.5).

Modes of Movement

There were more than 40,000 vehicles registered with Regional Transport Office (RTO), Meerut, on 31st of March, 1987. Of these 21,363 were two-wheelers, 11,680 tractors, 2,374 public and private carriers, 2,335 four-wheelers (cars and station wagon), 1,360 buses, 380 taxis, 301 jeeps, 140 three-wheelers, 84 three-wheeler carrier and rest other vehicles. Modes of transport are essential for the travelling up to the desired places. Intensity of these modes over various routes explains the relative gravity of the routes. Distance and amount of movements are closely associated with the modes of transport. Further, capacity to transport varies with the type of mode available and extent of route it covers. With the introduction of efficient modes of transport, physical distances between the places have been reduced considerably into time distances. Modes of transport have, thus, helped in maintaining space relations and bridging the missing links between the places (Table 4.7). In the absence of transport links and modes, nearest places may remain inaccessible. It is, therefore, essential to link the places and provide them with alternate modes of transport to increase the intensity of interaction. The common modes of transport in the region are train, buses (private and government), matador, taxi and tonga whereas private modes of transport are motorcycle, scooter, tractor, *buggi*, bicycle, etc. These modes are not available for the movement of people but are used as the personal mode of transport by the owner. Government buses and UPSRTC undertaking buses[1] are used as the modes of transport by maximum number of passengers. About 33 per cent of the total human movement was through these buses. Private buses, permitted on various routes, accounted for about 31 (30.79) per cent of the total traffic movement. Both the types of buses together accounted for carrying over 63 per cent of the total traffic. Matadors and three-wheelers are the short distance modes of interaction. These modes accounted for about 20 per cent of the total interaction (Fig. 4.6). Trains are very important modes of transport for long distance movements. Of the total interaction about four per cent was through railways. Taxi, car, motorcycle, scooter accounted for 3.42 per cent of the total interaction. This type of movement is relatively costlier and indicates the order of investment for the interaction. Being economically advanced region, share of these modes of interaction is likely to increase in future. *Tonga* serves as the most important mode of rural transport. Almost every village in the neighbouring rural areas has a *tonga* facility for the movement to the city. *Tonga* accounted for about three (3.08) per cent of the total traffic arrival to the city. With the advancement in agriculture, there has been a phenomenal increase in the number of tractors in rural areas. Besides being an important tool of agricultural operation, tractor also serves as the mode of rural transport. Tractors are used for carrying agricultural produce to the market and rural people to the city. Nearly 72,000 passengers have used tractors as their modes of transport. ***Buggi***, the cart, is a common goods carrier and easily available mode of transport in the rural areas. It acts as a mode of transport to the farmer for linking his place of residence with his agricultural fields. Sometimes, it also serves as a mode of transport to nearby villages. About 36,000 persons were reported to have used *buggi* as a mode of transport for their movements. Remaining 3.60 lakh persons have used other modes of transport. It is apparent from these facts that private modes of transport are gradually gaining significance over public modes of transport. While distant places have been connected through deluxe and super fast modes of transport, neighbouring places have been left inadequately served by the public modes of transport. As a result of this process, time distance has reduced considerably. Intensity of interaction decreases with the order of mode available. For increasing the pressure of interaction with Meerut city, frequent and alternate modes of transport need to be improved on all the routes. Both public as well as private modes of transport need to be improved for meeting the growing interaction demand of local as well as neighbouring areas.

TABLE 4.7: Meerut City — Modes of Human Movement, 1984

Modes	*Human movement (in numbers)*	*Proportion of movement through various modes*	*Number of sample schedules for field survey*
Roadways (UPSRTC)	1,72,08,471	32.71	50
Bus (Private permit)	1,62,00,000	30.79	40
Matador	1,08,00,000	20.53	30
Train	23,53,198	4.47	20
Bicycle	21,60,000	4.11	20
Car, Motorcycle, Scooter	18,00,000	3.42	20
Tonga	16,20,000	3.08	15
Tractor	72,000	0.14	10
Buggi	36,060	0.07	10
Others	3,60,000	0.68	10
Total movement	5,26,09,729	100.00	225

Source: Data for human movement through buses and trains have been collected from the office records of bus depots and railway stations whereas the amount of human movement through other eight modes have been collected through field survey.

Conclusion

The following conclusions emerge from the foregoing facts :

(i) About 20 (19.65) million people were registered on the move during 1984. As a result of this massive human interaction, Meerut city recorded an earning worth Rs. 106 million. With an interaction range of 1,600 km, Meerut recorded interaction with 203 places in different parts of the country.

(ii) Amount of human movement decreases with the distance and purpose becomes selective and of higher order with maintaining social relations and its proportion increases towards long distances. While most of the work-place interaction was through railways, most of the interaction for legal advice was through highways. Similarly interaction for purposes, like, social, political and administrative, was more through railways, whereas interaction for trade, alcoholic drinks and entertainment was relatively more through highways,

(iii) Amount of interaction is proportional to the frequency of movement. Similarly, ratio of interaction decreases with frequency of movement in case of highways. On the contrary, it increases in case of railways. Frequency distribution of movement presents declining trend in which frequency of daily movement formed peak of interaction, and

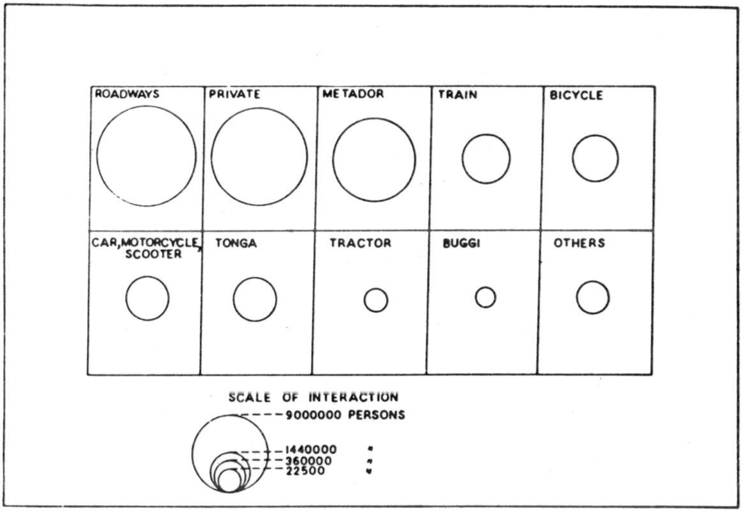

FIG. 4.6: Meerut city – Modes of conveyance used for interaction.

(iv) About 40,000 vehicles were registered as modes of transport with R.T.O., Meerut, on 31st March, 1987. Of these 21,363 were two-wheelers, 11,680 tractors, 2,374 carriers, 2,335 four-wheelers, 1,360 buses, 380 taxis and remaining about 500 private vehicles (like jeeps and two-wheelers). Buses are used as the modes of transport by maximum number of passengers. About 63 per cent of the total human movement was through these buses. Trains are the most important modes of transport for long distance movements. Nearly four per cent of the interaction was through railways and remaining traffic flow is through private modes of transport. Intensity of interaction decreases with the order of mode available. For increasing pressure of interaction with Meerut city, frequent and alternate modes of transport need to be improved on all routes.

NOTE

1. Movement through undertaking buses have been banned recently following a decision to this effect by the Court.

5

CORRELATES OF COMMODITY FLOWS AND HUMAN MOVEMENTS

INTRODUCTION

THIS CHAPTER is devoted to the study of factor correlates of commodity flows and human movements. Production, demand, price and distance were used to explain the flow of selected commodities whereas distance purpose, frequency and modes were used to explain the amount of human movements. The extent of association, its significance and the level of influence have been worked out to analyse the relative importance of the correlates. The correlates of prices and arrivals were calculated in cases of 10 commodities, viz., foodgrains, pulses, oilseeds, vegetables, spices, fruits, sweets, timber, animal products and livestock feed. To evaluate the response of demand on the supplies, six commodities were selected, viz., food items, sugar, coal, cement, cloth and kerosene. Apart from demand response, production response was also calculated. A set of five commodities, viz., agricultural produce, sports goods, cutlery, milk and textiles, were selected to analyse the impact of production of the outflow from Meerut city. The amount of human movement through railways, highways and through both the ways were discussed for 11 distance slabs. Similarly, frequency of human movements (six) included seven distance slabs for their explanation. On the basis of these results, three orders of pressure zones were identified. Since human interactions are purpose-oriented, movement for various purposes were tested over short, medium and long distances. Modes of conveyance are necessary for the movement of passengers and commodities over various distances. As many as 11 modes of conveyance were used to explain the relative importance of various modes and to assess the impact of distance (seven slabs) on these modes.

Methods used in this chapter relate to the working out of the index values, means, standard deviations, correlation and regression values for each set of variables. Index values have been calculated to remove the bias, vastness and incomparability of the data. Taking 100 as the index for the base year (1981), values for successive years were calculated. Separate programmes were framed for each exercises. These programmes were processed through computer (at Computer Centre, Council for Social Development, New Delhi) to ensure the accuracy of results. The entire chapter is arranged into eight exercises. Maps and graphs have been prepared to explain the trends of variables and their space relations.

Market Price Response

Market mechanism is the chief channel that connects producer with the consumer. Buyers and sellers both are influenced by the prevailing market prices. As such, prices tend to exert high influence upon various factors including arrival. This exercise deals with the price response on *mandi* arrivals. Market prices for selected 10 commodities were considered independent factors and weight of arrival for each of these commodities as dependent factors. Data used in the exercise refer to the periods 1981, 1982, 1983 and 1984. Correlations and multiple regression coefficients were calculated for the purpose (Table 5.1).

Results reveal that the behaviour of market prices and arrival of commodities varies significantly explaining limited response over each other. Separate accounts of selected *mandi* arrivals and their prices are given below:

Foodgrains

During 1981 the quantity of foodgrains arrival in the *mandi* was 3,850 tonnes while market price per tonne of foodgrains (average for the prices of all the foodgrains) was Rs. 1,942.30. The trend of foodgrains arrival registered rise during 1981-82, a nominal fall during 1982-83 and further rise during 1983-84. Trends of foodgrains prices, on the

TABLE 5.1: Trends of Mandi Arrival and Market Prices

Commodities	Arrival 1981		Index values (Based upon the index of 1981 - weight 100 and rate 100) 1982		1983		1984		Weight and rate relationship (r values)
	Weight in tonnes	Rate in rupees	Weight	Rate	Weight	Rate	Weight	Rate	
Foodgrains	3,850	1,942.30	117.48	119.40	87.22	117.03	94.88	109.11	+ 0.20
Pulses	258	3,337.50	227.13	101.87	1,350.00	119.85	1,360.08	101.12	+ 0.40
Oilseeds	856	4,500.00	143.46	103.33	5.37	122.22	152.57	97.22	- 0.80
Vegetables	574	1,143.20	108.19	215.69	116.38	193.60	141.46	181.85	+ 0.60
Spices	40	9,500.00	132.50	131.58	20.00	97.37	47.50	136.84	+ 0.40
Fruits	190	3,102.90	109.47	106.92	97.89	97.82	167.89	106.92	+ 0.90
Sweets	19,108	2,400.00	116.49	114.58	114.80	156.25	97.10	125.00	+ 1.00
Timber	8,100	600.00	138.72	116.67	149.70	116.67	267.18	133.33	+ 0.90
Animal Products	469	22,000.00	27.08	90.91	86.99	113.64	98.29	113.64	+ 0.10
Livestock feed	8,282	500.00	98.04	160.00	70.39	80.00	100.20	100.00	+ 0.60

other, registered substantial rise in the first year and subsequent fall in two successive years. Despite substantial rise in foodgrains arrival, prices tend to observe a little change. Arrival has inverse relationship with the distance. Price fails to influence the arrival significantly. Limited facilities of storage and an urgent need for the sale of the product accelerate the arrival of foodgrains to the market.

Pulses

Meerut *mandi* registered a unique trend of increasing arrival of pulses. It needs to be tested as to what has been the price response to such a high arrival. Arrival of pulses to the market was 258 tonnes during 1981. It recorded sharp rise in the quantity of arrival with the index number moving up to 227 for 1982 and 1,360 for 1984. In response to such a high arrival, prices registered an insignificant increase over these years. Price index for pulses was 100 for the base year (1981). 101.87 for 1982, 119 for 1983 and 101.12 for 1984. Behaviour of price appears to be quite different in explaining the increasing arrival of pulses.

Oilseeds

During 1981 *mandi* recorded 856 tonnes of oilseeds as arrival with the prevailing market price as Rs. 4,500 per tonne. The index values for the quantity of arrival and prices were 143 and 103 for 1982, five and 122 for 1983 and 152 and 97 for 1984. Thus, fall in the amount of arrival caused corresponding rise in the market prices. As such, there is an inverse relationship between the arrival and the prices in case of oilseeds.

Vegetables

The arrival of vegetables amounted to 574 tonnes during 1981. The average market price for the period was Rs. 1,143 per tonne. There has been an overall increase in the index values of the prices and weight of the arrival. From 100 in 1981, price index reached to the level of 182 and that of arrival 141 by 1984. Despite overall increase in the index values of prices and arrival, the two trends are dissimilar. The period 1981-82 observed substantial increase in the vegetable prices, i.e., from 100 to 216, when increase in the quantity of arrival was marginal, i.e., from 100 to 108. Further, vegetable prices registered a downward trend, while their arrival registered an upward trend during 1982-1984. Thus, there is a higher variability in the vegetable prices compared to its arrival. The two factors have an inverse correspondence due to dissimilar trends. Here increase in the vegetable prices fails to accelerate the arrival

Spices

Arrival of spices in the market was 40 tonnes during 1981. For the same period market price per tonne of weight was Rs. 9,500. Over the years, arrival of spices registered a declining trend. From an index value of 100 in 1981, the arrival fell down to the index value of 47 and that of prices to the value of 136 by 1984. Unlike vegetables, spices present higher variations in arrival compared to their prices. The trend of arrival and prices registered almost the same scale of increase during 1981-82. It is followed by the considerable decline in the quantity of arrival and moderate decline in the amount of prices during 1982-83. Finally 1983-84 registered significant increase in the two attributes. However, increased index value was considerably higher variation compared to the prices. Consequently, the two attributes have low level of positive response.

Fruits

Arrival of fruits in the market amounted to 191 tonnes in 1981. Ever since, there has been an increase in the quantity of fruit arrival. As a result, index value reached the level of 167 by 1984 as against 100 in 1981. Increase in the amount of arrival was, however, maximum during 1983-84. Average price per tonne of weight (fruit) was Rs. 3,102 in 1981 which increased gradually in subsequent years. The price index rose from 100 to 106 during 1981-1984. It is evident that a nominal increase in the fruit prices exerts significant influence on the fruit arrival. Since there are high variations in fruit arrival as compared to their prices, relationship between the two becomes positive though of low level.

Sweets

Arrival of sweets in the market was about 19,000 tonnes during 1981.

Quantity of arrival observed substantial increase during 1981-82 and gradual decline during 1982-1984. On the other, the price index recorded regular increase from 1981 up to 1983. Price per tonne of sweet was Rs. 2,400 during 1981. Within a short span of two years price index reached the level of 156 from 100 in 1981. Despite considerable increase in the sweet rates during 1981-83, amount of sweet arrival witnessed nominal increase. Price index fell down to 125 during 1983-84. Thus, there is negative correlation between the prices and arrivals of sweet in the market.

Timber

Timber presents a unique trend as arrival and prices both registered a rising trend showing little variations. Timber arrival in the city was 8,100 tonnes in 1981. The index of timber arrival reached the level of 267 by 1984 from 100 in 1981. Similarly, timber prices registered regular increase during this period. The average price per tonne of timber was Rs. 600 during 1981. The price index reached the level of 133 by 1984 from 100 in 1981. Obviously, there is a positive relationship between the two attributes. Degree of association between the two is significant as nominal increase in the price index causes significant increase in the quantity of arrival. It has, however, been observed that variation has become wider over the years. As such, it makes clear that increasing timber arrival in the city is mainly due to the growing local and regional demand for timber. Timber prices, on the other, tend to increase gradually, partly due to inflation and partly due to increasing cost of transport and other factors from the place of origin.

Animal Products

Animal products recorded an arrival of 470 tonnes of weight during 1981. Weight of arrival declined sharply during 1981-82, the index of arrival being 27 in 1982 as against 100 in 1981. The index of arrival increased subsequently but it could not attain the level of the base year, i.e., 98 in 1984 whereas 100 in 1981. Market price of animal products was Rs. 22,000 per tonne during 1981. Price index recorded the little variations in the course of time. It rose to the level of 114 by 1984 from 100 in 1981. With exception to a nominal decline in the price index (91) during 1982, there has been a regular increase until 1984.

The profile of temporal data suggests a positive response between the two attributes. Further, it is important to note that price and weight relationship is highly significant in this case compared to the fruits, timber and pulses.

Livestock Feed

During 1981 the average market rate per tonne of livestock feed was Rs. 500 whereas the quantity of arrival in the *mandi* was about 8,282 tonnes. There were significant changes in the prices and arrivals during 1981-84. While quantity of arrival registered regular decline up to 1983, (i.e., from 100 index value for 1981 to 70 index value for 1983) prices registered irregular trend of steep rise during 1981-82 and steep fall during 1982-83. It is interesting to note that despite marked variations in the intervening period, the two trends end (1984) almost at the same point of index (Fig. 5.1). As a result of the irregular behaviour of the two trends relationship turns to be negative.

Influence of Market Prices on Arrival of the Commodities

Having analysed the temporal patterns of commodity arrival and their market rates, it was thought desirable to work out extent of correlation and to measure the influence of rates on *mandi* arrivals. The exercise reveals that there was an inverse relationship between the two attributes in 1982, the value of correlation coefficient being –0.092 and that of regression coefficient 0.0085. The analysis further adds that per unit increase in the prices adversely influenced arrival by 0.12 unit. Responses between the two attributes became positive, though insignificant, by 1983. The value of correlation coefficient was 0.023 and that of regression coefficient 0.005. Per unit change in the prices influenced arrival by 0.29 unit. The level of response between the two attributes increased substantially by 1984. There was a negative correlation of the order of –0.249 and regression coefficient of the level of 0.062. Per unit increase in the prices, adversely influenced the arrival by 3.835 units.

Market prices for the past four years explain that they exert limited influence on *mandi* arrivals. However, the degree of relationship and its significance increased over time. Such a situation, in the result, may be explained that excess arrival of a commodity in the market pulls down the rates whereas limited arrival of a commodity pushes up the rates. Hypothesis that increase in the market prices positively influences the quantity of arrival does not hold true in the present case. As such market rates have limited role in influencing the quantity of arrival (Table 5.2).

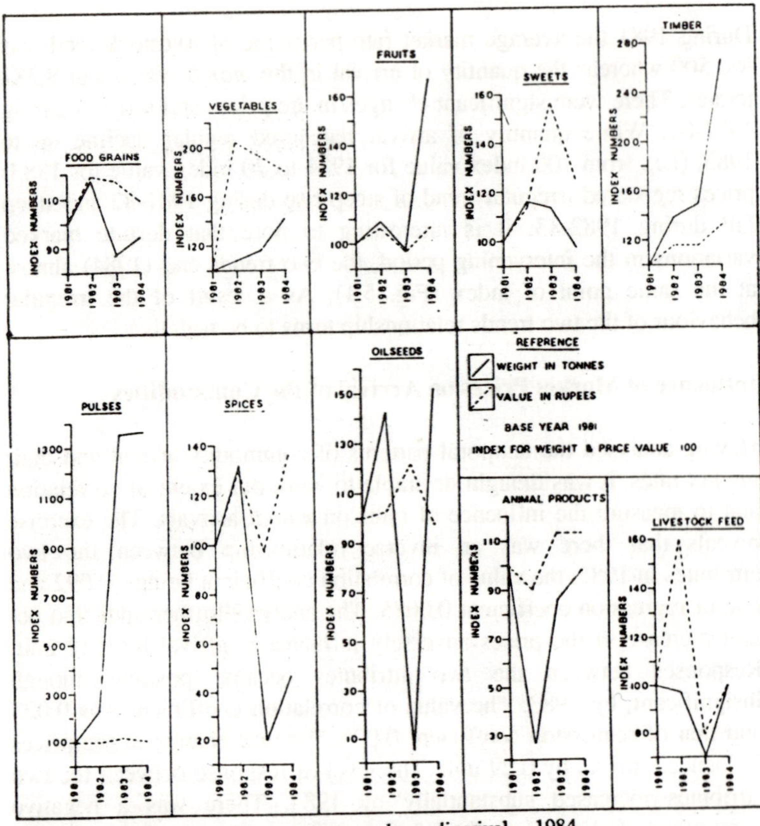

FIG. 5.1: Meerut city – Price response and *mandi* arrival – 1984.

Demand Response

This exercise deals with the analysis of demand and supply of selected commodities. Demand and supply influence each other to a large extent. Supply (dependent variable) has been considered dependent on the market demand (independent variable). Exercise refers to the periods of 1981, 1982, 1983 and 1984. Six commodities, viz., food items, sugar, coal, cement, cloth and kerosene oil, have been selected to work out this relationship. Demands of these commodities were calculated with the help of data of civil supplies through fair price shops and flow differentials in case of Meerut city. On the assumption that supply of these commodities through Fair Price Shops meets only the partial demand of the city, rest of the demand is fulfilled through flow differentials. Demand is, as such, taken to be the product of the city supplies and the flow differentials in the city. To explain the prevailing trends of demand and supply, separate accounts are given for each commodity (Table 5.3).

TABLE 5.2: Correlates of Price Response

(A) Correlation coefficient

	W82	*W83*	*W84*	*R82*	*R83*	*R84*
W82	1.00000	0.73394	0.77187	–0.09255	0.02638	–0.14582
W83		1.00000	0.99156	–0.21102	0.02352	–0.22322
W84			1.00000	–0.23813	–0.00165	–0.24981
R82				1.00000	0.50045	0.74166
R83					1.00000	0.72713
R84						1.00000

(B) Regression coefficient

	R^2	*B*	*Beta*
Dependent variable = W82	–0.00857	–0.12404	–0.09255
Variable R82 (Constant)		137.49631	
Dependent variable = W83	0.00055	0.29358	0.02352
Variable R83 (Constant)		174.22064	
Dependent variable = W84	–0.06241	–3.83574	–0.24981
Variable R84 (Constant)		714.93318	

Note: W = stands for weight of the commodity and, R = stands for rate of the commodity.

Food Items

Demand of food items was of the order of about 67,000 tonnes in Meerut city during 1981. Demand increased to the level of 515 thousand tonnes by 1983. A sharp increase in the demand was registered during the period 1981-83. There has been a slight decline in the demand of food items during 1984. A demand worth about 300 thousand tonnes of food items was recorded during 1984 compared to its demand of 515 thousand tonnes a year ago. Supply of food items presented positive response to the demand. It is important to note that supply (73,000 tonnes) exceeded the demand during 1981. The trend of demand and supply registered almost the same level in successive years, though supply did not exceed the demand as observed in the very first year. It holds true that the increase in the demand of a commodity causes corresponding increase in the supply of food items.

Sugar

Supply of sugar through fair price shops was about 7,000 tonnes as against its demand of about 12,000 tonnes in 1981. Supply of sugar increased gradually whereas demand of sugar increased sharply causing greater variability in demand than its supply over the years. The supply and demand differential explains a gap in the entire trend. This gap widens along the time. It is evident that supply of sugar increases with the increase in its demand over time. The two attributes are positively associated.

Coal

Coal is supplied to the city from the States of Bihar and Madhya Pradesh. Due to growing industrial and domestic demands of coal, city registered huge quantity of coal supply every year. About 92,000 tonnes of coal were supplied to the city as against its demand of 96,000 tonnes during 1981. Supply increased to 98,000 tonnes against its demand of 100 thousand tonnes by 1984. In terms of variation, demand presents relatively higher variations compared to the supply. Such a situation arises possibly because of the checks posed by the coal producing States. With the result, supply grows at a normal scale irrespective of sharp rise in the coal demand due to urban industrial growth of the city.

Cement

Alike coal, cement is another non-regional product and is supplied to the city from the States of Madhya Pradesh, Rajasthan and eastern Uttar Pradesh. Demand for the cement was 117 thousand tonnes during

1981, which fell down subsequently to about 30,000 tonnes by the year 1984. Decline in the cement demand was partly due to the directed supply of the cement to New Delhi for the construction of number of sports stadia and fly-overs on the eve of the IX Asian Games in New Delhi during 1984. With the result, construction works suffered and had to switch-over to other alternatives. Supply of cement to the city was reduced considerably. Meerut city recorded 113 thousand tonnes of cement supply during 1981, which fell down to the level of about 25,000 tonnes by the end of 1984. Both supply and demand registered high variations in case of cement. Demand is, therefore, not only a local factor but a regional factor as well. Regional demand conditions influenced the supply at all points in the region. Further, metropolitan demand always affects the amount of supply in neighbouring cities and towns. Demand is, as such, a variable influenced and organised over the space in a hierarchic system of settlements.

Cloth

Cloth is one of the basic human needs and is always in ever-increasing demand after food. Being an important centre of textile production, Meerut exports cotton textile to the various parts of the country and also abroad. Demand of various types of cloth in the city was estimated as about 28,000 tonnes for 1981. Demand increased to the level of 32,000 tonnes by 1984. There has been a regular increase in the cloth demand. On the contrary, supply of cloth was only about 12,000 tonnes in 1981 which rose to about 18,000 tonnes by 1984. Unlike demand, supply did not register a regular increase. Instead, supply registered decline during 1981-82. There are marked variations in supply and demand curves of cloth. Woollen, silk, synthetic and super fine cloths and *sarees* are supplied to the city from other parts of the country. Supply and demand seem to be influenced greatly by the textile production in the city. The two trends present positive association. However, the level of relationship is low and also insignificant. Thus, the demand of cloth provides limited explanation to its supply. Demand of cloth is largely influenced by the purchasing power of the local people.

Kerosene

Kerosene is one of the non-regional products and is supplied to the city from the States of Assam and Gujarat. There is a great demand for kerosene in Meerut city. While demand grows regularly, supply tends to be irregular. Meerut city registered a demand of about 29,000 tonnes of kerosene during 1981. Subsequently, the demand rose to 31,000 tonnes by 1982-83 and 30,000 tonnes by 1983-84. Demand is characterised by an extremely low variations and slow growth. As against the above demands, supply of kerosene was 23,000 tonnes in 1981 and 22,000 tonnes in 1984. Therefore, supply of kerosene continues to remain roughly the same. Consequently, gap between demand and supply continues to widen. The two factors are positively associated. Demand and supply differential continues to increase and affect adversely the degree of relationships. As such rise in the demand may not increase the supply in the same proportion (Table 5.3).

Nature of relationship between demand and supply varies significantly from one commodity to the other. While factors correspond positively, extent of its relationship is also influenced by the local factors, distance, transport cost and, above all, the availability of a commodity.

Impact of Demand on Supply

To test the effect of demand on the supply, correlation and regression exercises have been worked out. Table 5.4 reveals that demand and supply have extremely high positive correlation. The correlation values ranged between 0.73 (kerosene) to 0.99 (food items). While cement, coal and sugar had more than 0.95 correlation each, cloth recorded a correlation of 0.76. Regression coefficient for food items was 0.973 and for sugar 0.926 whereas it was considerably low in case of coal, cement, cloth and kerosene. The respective regression coefficients were calculated as 0.010, 0.009, 0.092 and 0.009 respectively. These results explain that the influence of demand on supply of food items and sugar has been close to one. That means the per unit change in the demand affects approximately the same unit of supply. In sharp contrast to the above generalisations, the demand of remaining four commodities regulates the limited supply. Per unit change in the demand affects supply 1/10th in case of coal and cloth whereas it affects 1/100th in case of kerosene and cement.

TABLE 5.3: Nature of Demand and Outflow

(*Weight: In Tonnes*)

Commodities	*1981*		*1982*		*1983*		*1984*	
	Demand	*Outflow*	*Demand*	*Outflow*	*Demand*	*Outflow*	*Demand*	*Outflow*
Food items	67,425.12	73,535.43	3,30,813.01	3,30,593.63	5,15,122.51	5,09,309.01	3,05,498.70	3,04,533.42
Sugar	11,847.40	6,958.20	24,028.70	18,593.70	20,700.46	15,084.80	20,907.66	14,383.90
Coal	95,932.00	93,798.00	90,028.40	87,650.00	94,606.40	89,768.00	1,00,000.10	97,700.00
Cement	1,17,151.20	1,13,868.00	1,04,723.20	97,320.00	1,18,322.85	1,09,610.00	30,133.86	24,737.00
Cloth	28,480.07	1,250.92	25,896.71	940.97	31,566.17	1,155.22	32,374.89	1,790.83
Kerosene	29,433.50	22,895.00	31,287.80	24,425.00	31,068.20	23,813.00	30,126.40	21,808.00

TABLE 5.4: Correlates of Demand Response

(A) Correlation coefficient

	VAR007	*VAR008*	*VAR009*	*VAR010*	*VAR011*	*VAR012*
VAR001	0.99999					
VAR002		0.99339				
VAR003			0.95784			
VAR004				0.99844		
VAR005					0.76276	
VAR006						0.73466

Note: Other values of the correlation coefficients had no logical base hence were left unrecorded.

(B) Regression coefficient

	R^2	*B*	*Beta*
Dependent variable = VAR007	0.99999	0.97354	0.99999
VAR001 (Constant)		784159.58679	
Dependent variable = VAR008	0.98683	0.92697	0.99339
VAR002 (Constant)		–420126.17847	
Dependent variable = VAR009	0.91746	0.01037	0.95784
VAR003 (Constant)		–6463.03510	
Dependent variable = VAR004	0.99688	0.00989	0.99844
VAR010 (Constant)		–5192.20811	
Dependent variable = VAR005	0.58180	0.09274	0.76276
VAR011 (Constant)		–145868.45305	
Dependent variable = VAR006	0.53973	0.00974	0.73466
VAR012 (Constant)		–6459.99790	

Demand variables	*Supply variables*
VAR001 = Food items	VAR007 = Food items
VAR002 = Sugar	VAR008 = Sugar
VAR003 = Coal	VAR009 = Coal
VAR004 = Cement	VAR010 = Cement
VAR005 = Cloth	VAR011 = Cloth
VAR006 = Kerosene	VAR012 = Kerosene

It is, thus, evident that low level of demand has been recorded in cases of six commodities. Response of demand was extremely high in case of food items and sugar, medium in case of coal and cloth and low in case of cement and kerosene. Further, commodities which are produced locally respond to the demand in the same scale. Since cloth is produced locally and is supplied outside in a large scale, its demand remains low to moderate. On the contrary, commodities like coal, cement and kerosene are not produced locally and depend entirely upon the supply from outside. Consequently, cement and kerosene have relatively low response and coal has moderate response to the demand.

Production Response

This exercise deals with the production and its influence on the outflow of commodities. To analyse the response of production, five commodities, viz., milk and milk-made items, textiles, sports goods, agricultural produce and cutlery goods, have been selected and data were obtained for the periods 1981 to 1984. Productions of these commodities were considered independent variables and the amount of their outflows dependent variables for this exercise. With the help of the above data, exercises on correlation and regression have been worked out. Results reveal significant variations in the production and outflow trends of these commodities. Behaviour of data and the group response of each commodity are discussed below separately (Table 5.5).

Agricultural Produce

According to 1981, agricultural production in the district was 8.2 million tonnes which increased to 8.9 million tonnes by 1984. There has been a regular increase in agricultural production with the exception to the period 1983-84. Of the total agricultural production, about 8.5 thousand tonnes was recorded as outflow from Meerut in 1981. There has been a considerable decline in the amount of outflow in subsequent years. The amount of outflow was 3.95, 1.99 and 1.20 thousand tonnes for the respective periods of 1982, 1983 and 1984. Thus, there is an inverse relationship between the production and the outflow. Decline in agricultural outflow is partly due to the increased grain production in the country and partly due to the increased local consumption and the increased facility of storage.

TABLE 5.5: Trends of Production and Outflow

(Weight: In Tonnes)

Commodities	1981		1982		1983		1984	
	Production	*Outflow*	*Production*	*Outflow*	*Production*	*Outflow*	*Production*	*Outflow*
Agricultural produce	82,63,838.00	8,052.00	86,92,361.00	3,952.00	92,41,713.60	1998.00	89,23,111.30	1,209.00
Sports goods	2,178.71	2,074.96	2,293.41	2,163.59	2,515.81	2329.45	2,815.59	2,559.63
Cutlery goods	76.27	72.64	73.06	68.92	92.21	85.38	95.73	87.03
Milk	24,137.01	11,015.36	18,455.14	14,550.34	17,601.45	14,875.85	20,829.42	11,668.35
Textiles	31,678.42	29,745.00	28,728.43	26,849.00	35,343.00	32,725.00	35,760.00	34,170.00

Sports Goods

Sports goods registered regular increase in the production over time. While it was about 2,200 tonnes in 1981, it rose to the level of 2.8 thousand tonnes by 1984. Since production of sports goods increased at a faster rate, there has been a phenomenal increase in its outflow. With the result production and outflow trends move parallel to each other. In other words, it explains a high degree of positive association between the two attributes. Since there are only a few centres of sports goods production in the country, huge market demand accelerates supply of sports goods from Meerut. Being qualitatively superior, sports goods of Meerut enjoy virtual monopoly in the national market. As a result of the established credit in the market, production and outflow increase simultaneously. Production, therefore, has corresponding influence on its outflow.

Cutlery Goods

Cutlery is one of the traditional industries of Meerut city. Cutlery is one of the famous industrial products of Meerut. Cutlery production is largely based on the traditional technology and is organised at the level of household and small scale industrial units. Production of cutlery goods was about 76 (76.27) tonnes in 1981 which increased to about 96 (95.73) tonnes by 1984. There has been marked variations in the annual production. Production of cutlery goods registered a nominal fall of three tonnes during 1981-82. This was largely due to the then prevailing disturbances (communal) in the city. There has been a regular increase in the production since 1982. Outflow follows the identical trend of production, though there has been minor variations in the two trends. It is, thus, clear that despite low production, established quality of the product always keeps it in the ever increasing demand. The amount of cutlery production and its outflow have high degree of positive association. The amount of production exerts corresponding influence upon the outflow. Such a response is due to the limited competition of the product. Because of the unorganised industrial sector, level of the production and outflow could not be raised significantly.

Milk

Meerut is known to have experienced white revolution caused by the planned development in milk production. There has been an overall increase in the milk collection and supply. However, due to the development of private chilling plants and direct supply of milk to Delhi, production of milk and milk-made items of F.B.D., Partapur (Meerut) registered uneven production over the years. While milk production was about 24 (24.13) thousand tonnes in 1981, which fell down to about 21 (20.82) thousand tonnes by 1984. Meerut city registered about 11,000 tonnes of milk outflow in 1981. It increased the level of outflow at 12,000 tonnes by 1984. Like milk production, its outflow registered a regular increase between 1981 to 1983. Outflow of milk recorded a decline of about 3,000 tonnes during 1983-84. Thus, two factors, production and outflow, respond positively in case of milk as well. However, there is an inverse relationship between the factors at micro-level. Despite nominal decline in the production, outflow registered a gradual increase leading to such a relationship at the micro-level. The gap between the production and outflow has, however, reduced considerably over time. Growing demand for milk has been responsible for such a reduction.

Textiles

Alike sports goods and cutlery, Meerut is also known for its textile production. Production of handloom cloth ranged between 32,000 tonnes in 1981 to 36,000 tonnes in 1984. Textile outflow from Meerut was nearly 30,000 tonnes in 1981 which rose to the level of 34,000 tonnes by 1984. Figure 5.2 shows parallel trends of textile production and outflow showing little gap between the two trends. As such, there is a high positive association and influence between the production and outflow of textiles.

Effect of Production on Outflow

To test the influence of production on outflow, correlation and regression exercises have been worked out. The foregoing details reveal that sports goods (0.999), cutlery (0.997) and textiles (0.991) have extremely high positive correlation whereas agricultural products (–0.903) and milk (–0.939) have high negative correlation with their respective outflows. While production of milk and agricultural products present inverse relationship, production of industrial goods present positive association with their outflow. Some amount of the

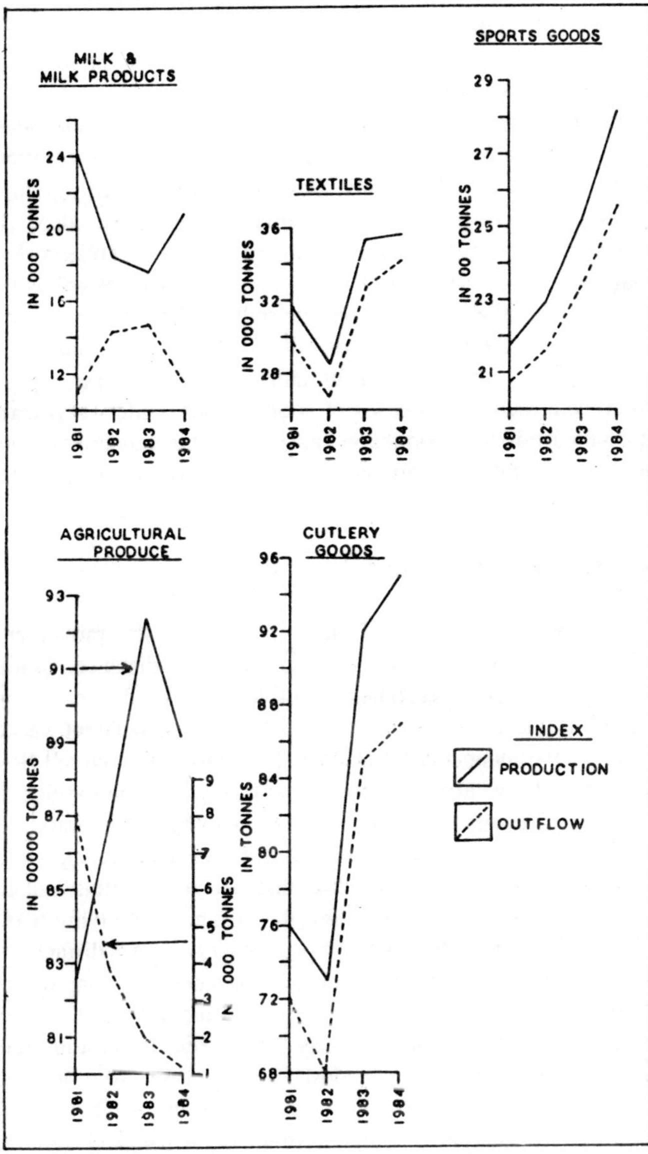

FIG. 5.2: Meerut city — Production response to outflow — 1984.

agricultural products and milk is processed before being supplied out. As a result amount of outflow remains much less than its production. Scope of variation between production and outflow of primary products remains high because of the increasing local consumption and processing into mill products. Variability appears to be the main factor behind inverse relationship (Table 5.6).

Regression coefficient between agricultural production and its outflow reveals that per unit increase in the production adversely affects the outflow by 1,215 units. Despite increase in the agricultural production, outflow remain low. Similar results have also been obtained in case of milk. Here amount of outflow increases despite marginal decrease in the level of milk production. Dichotomy is partly due to the competition posed by private milk chilling plants and partly due to increase in the production of milk-made items. Per unit of milk production deters outflow by 1.39 units. On the contrary, positive responses were recorded in cases of sports goods, cutlery and textiles. Per unit increase in the production causes simultaneous increase in the outflow by 1.32 units of sports goods, 1.24 units of cutlery and 101.11 units of textiles.

Distance and Flow of Commodities

Distance is one of the basic factors influencing the movement of commodities. Amount of flow decreases with the distance. Similarly number of commodities becomes selective with the distance. Supply and receipt of the commodities were considered dependent variables on distance – the independent variable. To measure the overall flow of commodities, inflow and outflow records of the commodities were added. Besides measuring the overall flow, number of commodities at various distance slabs was also calculated. Of the total receipts (Table 5.7), nearly 40 per cent was from short distance zone (within a distance of 500 km), 36 per cent from medium (the distance zone lying between 500 to 1,000 km) and remaining 24 per cent from long distance zone (beyond 1,000 km). In terms of number of commodities, nine out of 16 arrivals were from short distance, four from medium and remaining three from long distance zones. Nearly 20 per cent of the total receipt was from the State of Uttar Pradesh, 15 per cent from the Union Territory of Delhi and four per cent from the States of Rajasthan and Punjab. States of Gujarat, Madhya Pradesh, Bihar and Jammu & Kashmir together shared nearly 36 per cent of the receipt. States of north-eastern and southern India shared about a quarter of the total receipts. Amount of inflow and number of commodities decrease with the distance.

TABLE 5.6: Correlates of Production Response

(A) Correlation coefficient

	VAR006	*VAR007*	*VAR008*	*VAR009*	*VAR010*
VAR001	–0.90358				
VAR002		0.99998			
VAR003			0.99727		
VAR004				–0.93950	
VAR005					0.99141

Note: Other values of the correlation coefficients had no logical base hence were left unrecorded.

(B) Regression coefficient

	R^2	*B*	*Beta*
Dependent variable = VAR001	0.81645	–1215.42577	–0.90358
VAR006 (Constant)		92424512.34965	
Dependent variable = VAR002	0.99996	1.31652	0.99998
VAR007 (Constant)		–55329.49094	
Dependent variable = VAR003	0.99455	1.24549	0.99727
VAR008 (Constant)		–1344.43767	
Dependent variable = VAR004	0.88266	–1.39600	–0.93950
VAR009 (Constant)		3844208.15091	
Dependent variable = VAR005	0.98290	101.11146	0.99141
VAR010 (Constant)		166207.83302	

Production variables	*Outflow variables*
VAR001 = Agricultural produce	VAR006 = Agricultural produce
VAR002 = Sports goods	VAR007 = Sports goods
VAR003 = Cutlery goods	VAR008 = Cutlery goods
VAR004 = Milk	VAR009 = Milk
VAR005 = Textiles	VAR010 = Textiles

Relationship between distance and outflow of commodities has been worked out to estimate the gap between the receipt and the supply and to discuss the causes for such a variation. Commodities weighing 48,000 tonnes were registered as outflow from Meerut during 1984. Of the total outflow (supply), about 60 per cent was for the places within short distance, about six per cent for the places lying in the medium distance and remaining 33 per cent for the places lying in the long distance zones. Result is interesting as it explains relatively higher outflow towards long and short distance zones compared to medium distance zone. This is a bi-modal situation in which medium distance zone does not influence proportionate outflow. Sports goods, jaggery, printed books and defence goods are the chief items of outflow within short distance zone. Combination of the commodities ranged between two to eight in this zone. On the contrary, medium distance zone is characterised by the supply of printed books, fertilizer, miscellaneous goods and defence goods. It should be explained that Meerut is not a centre of fertilizer production but it is an important centre of fertilizer storage and distribution. Number of commodities supplied to this zone ranged between one to four. Long distance zone recorded supply of band goods, sports goods, khandsari molasses, jaggery, crushed bone and defence goods. Combination of commodities ranged from one to five in this case. These facts further add that function of distance decay fails to explain the amount of outflow and the selectivity of commodities. Goods and commodities move towards short and long distances bypassing the places lying in the medium distance zone.

TABLE 5.7: Distance and Flow of Commodities — 1984

Name of the place/State	*Distance of State capitals from Meerut city*	*Weight in tonnes*			
		Receipt of commodities	*Percentage share*	*Supply of commodities*	*Percent-age share*
Delhi	68	1,05,882	14.96	19,385	40.36
Haryana	156	—	—	76	0.16
Rajasthan	376	20,716	2.93	2,955	6.15
Punjab	415	9,748	1.38	2,130	4.43
Uttar Pradesh	461	1,42,236	20.09	4,111	8.56
Himachal Pradesh	565	—	—	35	0.07
Jammu & Kashmir	583	2,231	0.32	418	0.87
Madhya Pradesh	773	22,223	3.14	725	1.51
Gujarat	1,002	1,74,016	24.58	868	1.81
Bihar	1,002	61,389	8.67	1,273	2.65
West Bengal	1,455	80,203	11.33	9,435	19.64
Maharashtra	1,457	86,359	12.20	3,305	6.88
Orissa	2,129	829	0.12	2,686	5.59
Tamil Nadu	2,206	1,458	0.20	175	0.36
Assam	2,538	163	0.02	405	0.84
Kerala	3,150	461	0.06	47	0.10
Total		7,07,914	100.00	48,029	100.00

Effect of Distance on the Movement of Commodities

Associations between the distance and the movement of commodities have been worked out with the help of correlation coefficient. Movement is represented by the quantity of supply and the receipt of the commodities. Thus, distance and weight received and distance and weight moved have been used to work out the degree of dependence and association. Results reveal that distance and flow of commodities are inversely proportional. While coefficient of correlation between the distance and the receipt of the commodities was –0.43, correlation between distance and supply of commodities was –0.42. Further, correlation between the receipt and supply of commodities was +0.35. All these correlations are significant at all levels of significance. It is, thus, established that the amount of flow decreases with the corresponding increase in the distance. However, the two types of flows maintain positive relationship with each other (Table 5.8).

TABLE 5.8: Correlation Between the Distance and the Flow of Commodities

(A) Coefficient of correlation

	VAR001	*VAR002*	*VAR003*
VAR001	1.00000	–0.42558	–0.41666
VAR002		1.00000	0.35288
VAR003			1.00000

(B) Regression coefficient

	R^2	*B*	*Beta*
Dependent variable = VAR002	0.18112	–26.72232	–0.42558
VAR001 (Constant)		84187.68996	
Dependent variable = VAR003	0.17360	–2.32506	–0.41666
VAR001 (Constant)		6348.14191	

VAR001 = Distance in km
VAR002 = Receipts in tonnes
VAR003 = Supplies in tonnes

Regression coefficients reveal that per unit increase in the distance decreases about 2.7 units of the receipt of the commodities. Similarly, though not in the same proportion, per unit increase in the distance adversely affects the supply by 2.32 units. While regression coefficient of the receipt was 0.18, regression coefficient of the supply was 0.17. Beta values further add that the factor of distance explains receipt up to 42 per cent and that of supply up to 41 per cent. It is established that distance has a direct bearing on the flow of commodities. While receipt of commodities is explained to a greater extent supply of commodities is explained only to a limited extent. Similarly the values of the regression coefficients are low but they are significant and as such explain the results clearly.

Distance and Human Movement

Human movement is an important aspect of interaction over the space. This exercise deals with the function of the distance decay and analyses the influence of distance on the amount of human movement. Movements through railways and highways have been considered as dependent factors on the distance (independent factor). Results reveal that human movements, through railways, were maximum (77%) for short distance, moderate for long distance (about 19%) and the minimum for medium distance zones (4%). Thus, amount of human movement varies significantly from one zone to another. Figure 5.3 reveals that, in general, amount of human movement decreases with the distance. However, slab-wise movement does not confirm this view as decrease in the amount lies close the medium distance slab. This reality refers to the distortions present in the movement over varying distances. There is a significant negative correlation between the two factors. As such, human movement is adversely affected by increasing distance.

Movement, through highways, reveals slightly different result. Function of distance decay operates without distortions in this case. Of the total human movement in this case (172 million), nearly 60 per cent was for short distance, 37 per cent for medium distance and remaining three per cent for long distance. Thus, as compared to the railways, long distance movements are very limited through highways. Function of distance decay becomes relatively more explanatory in this case. There is a significant negative association between the two factors in this case explaining adverse impact of distance on the human movement (Table 5.9).

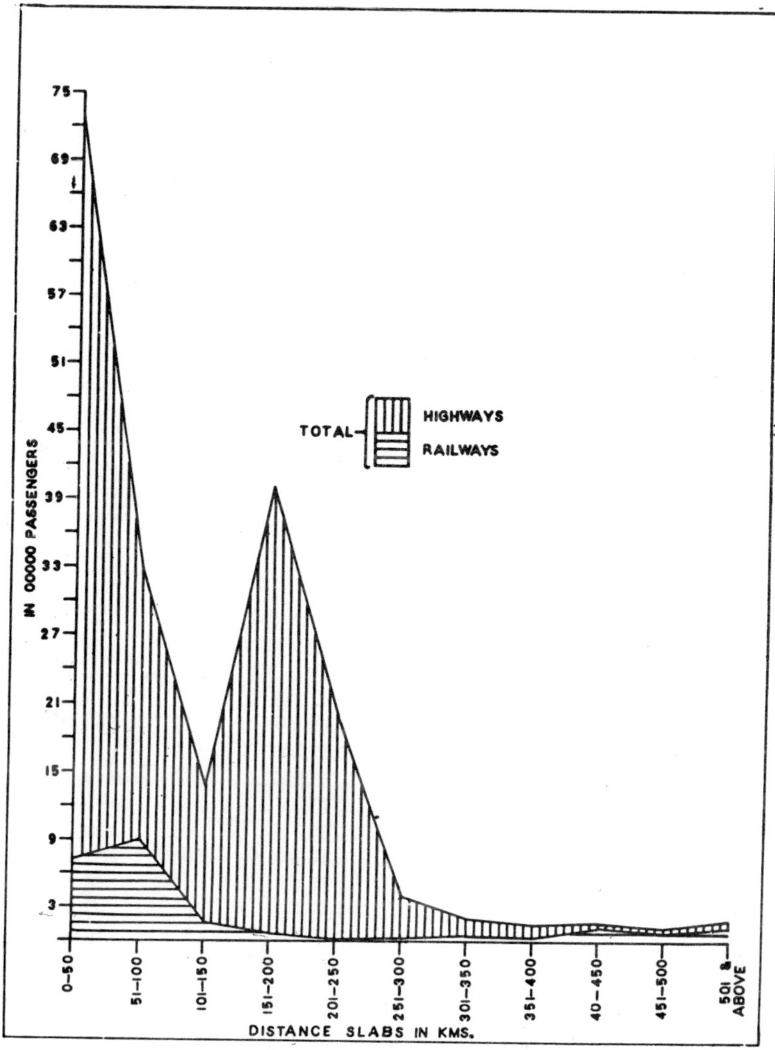

FIG. 5.3· Meerut city — Distance and human mobility — 1984.

Effect of Distance on Human Movements

Distance is one of the important factors influencing human movements. Distance has been grouped into short, medium and long distance slabs. Alike distance, human movements have been grouped into overall – through railways and through highways. Distance has been considered independent factor and human movements as dependent factor for this exercise. Results reveal that distance has a negative correlation with the amount of human movement. The value of correlation coefficient between the distance and the overall human movement was –0.80, between distance and the human movement through highways was –0.79 and between the distance and human movement through railways was –0.62. These results are significant at all levels of significance. Regression coefficients reveal that with per unit of distance overall human movement decreases by 11,326 units movement through highways by 10,156 units and movement through railways by 1,170 units. It is, therefore, clear that human movement follows the function of distance decay though it is more applicable in case of highways compared to railways (Table 5.10).

TABLE 5.9: Distance and Human Movement — 1984

Distance slab in km (From Meerut city)	*Amount of passengers movement through*				*Total*
	Highways	*Percent-age*	*Railways*	*Percent-age*	
0 - 50	66,39,219	38.58	7,42,286	31.54	73,81,505
51 - 100	24,51,595	14.25	9,06,504	38.52	33,58,099
101 - 150	12,80,644	7.44	1,69,786	7.22	14,50,430
151 - 200	39,60,254	23.01	58,896	2.50	40,19,150
201 - 250	20,43,793	11.88	13,997	0.60	20,57,790
251 - 300	3,79,925	2.21	24,290	1.03	4,04,215
301 - 350	2,04,036	1.18	47,780	2.03	2,51,816
351 - 400	1,35,350	0.79	31,458	1.34	1,66,808
401 - 450	29,002	0.17	1,38,472	5.88	1,67,474
451 - 500	31,782	0.18	85,205	3.62	1,16,987
501 and above	52,871	0.31	1,34,524	5.72	1,87,395
Total	1,72,08,471	100.00	23,53,198	100.00	1,95,61,669

Distance and Frequency of Movement

It has been established that human movement is influenced greatly by the factor of distance. However, it remains to be tested whether distance influences the frequency of visit. Frequency of visits to a place explains the dependencies among the places and their relative gravity. Interaction becomes more meaningful and important factor in explaining the development of a place if frequency of movement is higher. While occasional congregations leave temporary impression on the development of a place, regular movements of people leave long lasting impressions (permanent) on a place. Frequency of interaction is, therefore, important in explaining the nature of interaction within a given area. Frequency of movement along with the distance explains the relative order of the place interaction. This exercise deals with the intensity of interaction. Frequency of human movements and various distance slabs have been considered dependent and independent factors respectively. In all there are six frequency groups and seven distance slabs to explain this relationship. Frequency of visits and pressure of human movements are discussed below for understanding the intensity of interaction (Table 5.11).

TABLE 5.10: Nature of Correlation Between the Distance and the Human Movement

(A) Correlation coefficient

	VAR001	*VAR002*	*VAR003*	*VAR004*
VAR001	1.00000	0.62982	0.79497	0.80977
VAR002		1.00000	0.61033	0.69033
VAR003			1.00000	0.99444
VAR004				1.00000

(B) Regression coefficient

	R^2	*B*	*Beta*
Dependent variable = VAR002	0.39667	–1170.73473	–0.62982
VAR001 (Constant)		565147.50909	
Dependent variable = VAR003	0.63198	–10155.54236	–0.79497
VAR001 (Constant)		4611069.16364	
Dependent variable = VAR004	0.65573	–11326.27709	–0.80977
VAR001 (Constant)		5176216.67213	

VAR001 = Distance
VAR002 = Railway passengers
VAR003 = Roadways passengers
VAR004 = Total passengers.

TABLE 5.11: Distance and Frequency of Human Movement — 1984

Distance slabs (in km)	Cummulative frequency scale for the calendar year						
	Daily	*Monthly*	*Seasonally*	*Half-yearly*	*Yearly*	*Rare*	*Amount of movement*
1	2	3	4	5	6	7	8
Up to - 10	89,690	25,626	8,970	1,682	1,265	896	1,28,129
11 - 20	9,61,594	1,20,199	60,100	32,099	28,000	—	12,01,992
21 - 50	45,38,538	9,07,708	3,02,568	1,81,542	1,21,028	—	60,51,384
51 - 100	21,82,764	6,71,620	2,35,067	1,67,905	1,00,743	—	33,53,099
101 - 200	27,34,790	16,40,874	8,20,437	2,18,783	54,696	—	54,69,580
201 - 500	6,33,018	15,82,545	3,79,811	2,53,207	2,21,556	94,953	31,65,090
501 and above	11,244	65,588	46,849	37,478	18,740	7,496	1,87,395
Total	1,11,51,638	50,14,160	18,53,802	8,92,696	5,46,028	1,03,345	1,95,61,669
Percentage	57.01	25.63	9.48	4.56	2.79	0.53	100.00

Note: Projections are based on the net movement recorded for 1984 and field surveys (traffic) conducted during 1986. Various frequencies are, thus, estimated for the calendar year.

Frequency of Daily Movement

Of the total annual movements (1,95,60,000) about 57 per cent (1,11,50,000) are estimated to be commuters making daily movements. Frequency of daily movement varies significantly with the distance. For instance, of the total passengers making daily movements, nearly 41 per cent travel a distance of 50 km, 20 per cent a distance of 51 to 100 km, about 25 per cent a distance of 101 to 200 and remaining about 24 per cent a distance of over 200 km. Thus, frequency of daily movement forms a normal curve along the distance. Intensity of human movement remains high and of daily frequency up to a distance of 100 km from the city.

Frequency of Monthly Movement

About 26 per cent of the total human interaction was of monthly frequency. Frequency distribution reveals that 21 per cent of such movements belonged to a distance of 50 km, about 13 (13.39) per cent to a distance of 51 to 100 km and 33 per cent to a distance of 101 to 200 km. Remaining about one-third of the total movement originated from a distance beyond 200 km. It is thus, clear that movements of monthly frequency correspond largely with movements of daily frequency. It is, however, important to note that frequency of monthly movement is more marked with medium distance and has a normal distribution. Contrary to this, frequency of daily movement is largely associated with short distance and has acute distribution.

Frequency of Seasonal Movement

Thus type of movement accounted for about nine per cent (1.8 million passengers) of the total interaction. Maximum frequency (44.26%) of seasonal movement was recorded in the medium distance zone (101 to 200 km). Intensity of interaction decreases towards short and long distance zones. This type of movement was characterised by the large variations and irregular trend of distribution. Frequency of seasonal movement increases in three alternate steps. The first step with higher intensity of movement (16.32%) lay in the short distance zone between 21 to 50 km distance, the second step with maximum intensity (44.26%) lay in the medium distance zone between 101 to 200 km and the third step with high intensity of movement (20.49%) lay in the long distance zone between 201 to 500 km. It is, therefore, obvious that seasonal frequency of movement decreases towards short as well as long zones from the central (medium) zone. Since rate of decline is more marked towards short distance zone, relationship appears to be positive between the two.

Frequency of Half-Yearly Movement

About five per cent (8,92,000 passengers) of the total interaction belongs to the half-yearly movement. The medium distance zone recorded the maximum frequency (43.31%) followed by the long distance zone with moderate frequency (32.56%) and short distance zone with the minimum frequency (about 24%). Almost similar pressure of human movement was recorded in four out of seven distance zones. Remaining three slabs recorded uneven human movements. Due to limited variations in the distributions, frequency of half-yearly movement holds positive relationship with the distance.

Frequency of Yearly Movement

This type of movement accounted for about three per cent (5,46,000) of the total interaction. Frequency of yearly movement recorded maximum (44%) distribution in long distance zone and almost equal in short (27.53%) and medium distance zones (28.47%). The trend of the frequency distribution forms a bi-modal curve in which first peak of pressure lies within the distance slab 21 to 50 km (22.17 per cent of the movement) and the second between the distance zone 201 to 500 km. In the overall analysis, frequency of yearly movement increases with the distance. As such, it holds positive relationship with the distance. It is, therefore, very clear that annual movements are mostly for distant places up to 500 km.

Frequency of Rare Movement

This type of movement accounted for an insignificant proportion (0.53%) of the total interaction. Almost all interactions (nearly 99%) of this frequency belong to the long distance zone. On the contrary, short and medium range interactions were insignificant. Four out of seven distance slabs did not register any amount of such movement. It is, therefore, established that frequency of the rare movements hold positive correspondence with the distance (Fig. 5.4).

In the light of the above discussion, it may be said that the intensity of human movement decreases with the distance. Frequency of movement becomes selective and rare as one moves towards long distances. Frequency of daily and seasonal movement correspond closely with the short and medium distances whereas monthly and half-yearly movements correspond with the long and medium distances. Annual and rare movements are exclusively associated with the long distance.

Pressure Zones of Human Interaction

Pressure of human interaction along the distance marks the relative importance of slabs within each zone. On the basis of the distance covered and the number of passengers moved, pressure zones of human interactions were identified. The order of a zone depends upon the intensity of human movement recorded in that zone. Differentiation between the pressure zones is based upon the perceptible gap in the array of the distribution. Results reveal three distinct orders of pressure zones. A brief discussion on each zone is given below.

The First Order Pressure Zones: Table 5.11 reveals that maximum intensity (30.93%) of human movement was recorded in the distance slab of 21 to 50 km. It was closely followed by the distance slab of 101 to 200 km (27.96%). These two distance slabs are identified as the first order pressure zones. Most of the interaction in this zone is of daily frequency nature. However, intensity of daily frequency is relatively higher near the city. While pressure distribution seems to be concentrated near the city, it is relatively dispersed in other zones.

The Second Order Pressure Zones: There are two distance slabs in this zone having almost equal pressure of movement. These slabs are: (i) 51 to 100 km with a pressure of about 17 (17.17) per cent and (ii) 201 to 500 km with a pressure of 16 (16.18) per cent of the total movement. The first slab is characterised by the two-thirds of the total interaction as daily, one-fifth monthly and remaining about 15 per cent other frequency groups. As against this, second slab is characterised by half of its movement as monthly, one-fifth daily and remaining nearly 30 per cent other frequency groups. It is, thus, clear that while first slab has maximum pressure of daily movements, the second slab has maximum of monthly movements.

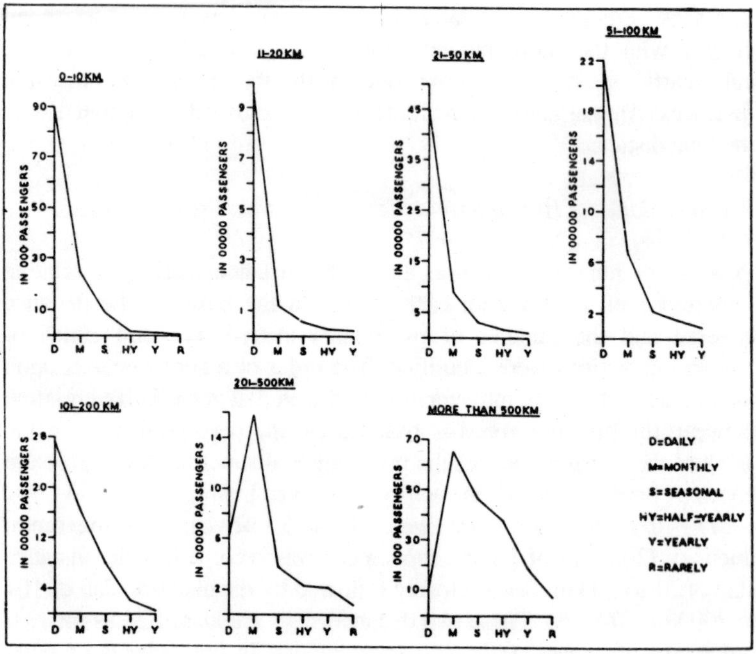

FIG. 5.4: Meerut city — Frequency of trips and movement of population — 1984.

The Third Order Pressure Zones: Low level of human movement characterised the third order pressure zone. Three distance slabs, viz., 11 to 20 km with a pressure of 6.14 per cent, over 500 km with a pressure close to one per cent (0.96%) and up to 10 km with a pressure of less than one per cent (0.66%) fall in this order. The first slab is characterised by 80 per cent of the total pressure of movement as daily, 10 per cent monthly and remaining 10 per cent of the rest of the frequencies. Movement is extremely concentrated in this slab and its dependency is maximum on the city. In terms of human interaction, this slab is the part of the city. The pattern of human movement varies significantly in the remaining two distance slabs. While second slab has more of monthly (35%), seasonal (25%), yearly (10%) and rare movements (4%), the third slab has relatively more of daily (70%) and monthly (20%) movements.

Meerut city, thus, presents a bi-variate picture of its pressure zones. The first, that it has established very close relations of interactions with its neighbouring areas. Second, it has increased its gravity of interaction over long distances. With the result, monthly, seasonal and annual movements (to and from distant places) have increased.

Impact of Distance on the Frequency of Human Movements

Frequency of human movement is an important factor in explaining the centrality of a place. Normally, distance deters the amount of movement which in its turn influences the frequency of movement. It is generally believed that the frequency of human movement decreases with the distance. But it remains to be investigated as to whether the same relationship holds true for all frequencies along all distance slabs. Contrary to the general understanding, frequencies of rare, yearly, half-yearly and monthly movements are positively associated with the distance. This kind of relationship is significant at all levels only for the frequency of rare movements. The correlation coefficient between the distance and the frequency of rare movements was 0.63, followed by the distance and the frequency of annual movement ($r = 0.35$). The distance and the frequency of half-yearly movement recorded a correlation of 0.25 whereas the distance and the frequency of monthly movement had a correlation value of 0.26. Inter-correlations between the above frequencies and with the frequency of daily movements were found positively correlated. However, degree of relationship was highly significant between daily, seasonal and half-yearly movements. The frequency of monthly movements recorded very high positive correlation with seasonal, half-yearly and yearly movements whereas frequency of rare movements recorded high positive correlation with the distance, yearly and seasonal movements. Contrary to the above, negative correlations were observed between the frequency of daily movement and the distance (*r* being 0.42), the amount of total movement and the distance (*r* being –0.17) and between the daily movements and rare movements (*r* being –0.30).

It is clear from Table 5.12 that frequency of human movement decreases with the distance. It is interesting to note that all frequencies, except daily, hold positive relationship with the distance. This response

TABLE 5.12: Correlation of Distance and Frequency of Human Movement

	VAR001	*VAR002*	*VAR003*	*VAR004*	*VAR005*	*VAR006*	*VAR007*	*VAR008*
VAR001	1.00000	0.16778	0.41590	0.25849	0.10576	0.25263	0.35287	0.62708
VAR002		1.00000	0.91249	0.78601	0.78809	0.82839	0.52575	0.02591
VAR003			1.00000	0.46470	0.52650	0.55478	0.28384	0.29656
VAR004				1.00000	0.90139	0.95537	0.72203	0.51885
VAR005					1.00000	0.79872	0.36303	0.15057
VAR006						1.00000	0.83243	0.51641
VAR007							1.00000	0.80230
VAR008								1.00000

VAR001 = Distance
VAR002 = Amount of movement
VAR003 = Frequency of daily movement
VAR004 = Frequency of monthly movement
VAR005 = Frequency of seasonal movement
VAR006 = Frequency of half-yearly movement
VAR007 = Frequency of yearly movement
VAR008 = Frequency of rare movement.

becomes more significant towards rare frequency of movements. Thus, while function of distance decay operates in case of daily movements, it fails to function in rest of the frequencies. Explanation to the problem may be that people from distant places cannot afford daily movements to the city. In addition to that people can afford to move over to distant places on their rare and seldom visits.

To test the influence of distance upon different frequencies of movement, regression coefficients have been worked out by assuming distance as independent and frequencies as dependent factors. Thus, there are in all eight factors, viz., seven dependent and one independent factors. High regression coefficients were recorded in case of rare (r = 0.39), daily (r = 0.17) and yearly (r = 0.13) frequencies. Per unit positive change in the distance affects daily movement adversely by 3,000 units. Unlike response with the daily frequency, distance exerts positive influence upon rest of the frequencies. Per unit positive increase in the distance affects monthly movements by 786 units, seasonal movement by 130 units, half-yearly movements by 112 units, yearly movements by 118 and rare movements by 97 units. It is, thus, established that distance exerts significant influence upon various frequencies of human movements. Amount of movement decreases with the distance despite maintaining positive relations with other than daily movements. Pressure of movement becomes thinner towards rare frequency. It affects the daily movements adversely. The amount of inverse change is quite large. Distance, thus, a vital role in limiting the extent of human movement and its frequencies (Table 5.13).

Distance and Purpose of Movement

Regression equations and coefficients of correlations have been worked out between the purpose of movement and the distance. The purposes included in the study are legal advice, entertainment, medical, business, alcoholic drinks, education, social, service, religious, search of job and political and administrative. These purposes are considered dependent factors. Distance included three slabs, viz., short, medium and long distance, and are considered as independent factors (Table 5.14).

Purpose of Movement

Of the total interaction (19.56 million), about one-third was for social purposes, 23 per cent for trade and business purposes, 21 per cent for legal advice and remaining for rest of the purposes. Social interaction was about 25 per cent for short distance zone, 70 per cent for medium distance zone and remaining about five per cent for long distance zone. In case of business interaction, medium distance accounted for 67 per cent, short distance for 32 per cent and long distance for less than one per cent. On the contrary, in case of interaction for legal advice short distance accounted for as much as 99 per cent and long distance for about one per cent. It is significant to note that medium distance zone recorded no interaction for legal advice. Most of the ineractions for political and administrative purposes was directed towards medium and long distance zones. Rest of the purposes recorded movements from short distance zones.

TABLE 5.13: Regression Coefficients of Distance and Frequency of Human Movement

	R^2	*B*	*Beta*
Dependent variable = VAR002	0.02815	–1764.49093	–0.16778
VAR001 (Constant)		3154984.43340	
Dependent variable = VAR003	0.17298	–3009.64596	–0.41590
VAR001 (Constant)		2207918.81774	
Dependent variable = VAR004	0.06682	786.30998	0.25849
VAR001 (Constant)		555676.67478	
Dependent variable = VAR005	0.01119	130.77575	0.10576
VAR001 (Constant)		238113.23920	
Dependent variable = VAR006	0.06382	112.11730	0.25263
VAR001 (Constant)		104624.03629	
Dependent variable = VAR007	0.12452	118.63503	0.35287
VAR001 (Constant)		53768.55739	
Dependent variable = VAR008	0.39323	97.31695	0.62708
VAR001 (Constant)		–5116.89200	

VAR001 = Distance
VAR002 = Amount of movement
VAR003 = Frequency of daily movement
VAR004 = Frequency of monthly movement
VAR005 = Frequency of seasonal movement
VAR006 = Frequency of half-yearly movement
VAR007 = Frequency of yearly movement
VAR008 = Frequency of rare movement.

TABLE 5.14: Distance and Purpose of Movement

Purpose	*Distance* — *Percentage share from the total distance slab* — *Short* — *Up to 100 km*	*Medium* — *101 – 400 km*	*Long* — *401 – 1600 km*	*Purpose-wise percentage from the total interaction*
Social interaction	14.85	52.96	80.70	32.54
Business and trade	13.80	36.14	11.67	23.42
Legal advice	38.50	—	4.77	21.27
Work place (service)	14.64	1.16	—	8.54
Medical care	6.20	—	—	3.39
Entertainment	6.20	—	—	3.39
Religious	0.42	5.73	—	2.71
Alcoholic drinks	3.09	—	—	1.70
Education	2.30	0.76	—	1.58
Search for jobs	—	3.10	—	1.34
Political and Administrative	—	0.15	2.86	0.12
Percentage total	100.00	100.00	100.00	100.00
Total interaction (Actual)	1,07,39,604	84,63,864	3,58,201	1,95,61,669
Percentage to total interaction	54.90	43.27	1.83	100.00

Of the total interaction nearly 55 per cent was for short distance, 43 per cent for medium distance and remaining two per cent for long distance. While short distance movements included maximum interaction for legal advice followed closely by social interactions, services and business interactions, movements for entertainment and medical aid accounted for six per cent each in short distance zones. About six per cent of the total short distance interaction was for drinks (alcoholic), education and religious purposes. Medium distance zone is characterised by the maximum movement for social purposes (52.46%) followed by business (36.14%). Interaction for religious purposes and for the search of job accounted for about nine per cent of total interaction. Minimum movement in this zone was for services, education and political and administrative purposes. About two per cent of the total interaction was related to these purposes. Long distance interaction was characterised by about 80 per cent of the movement for social purposes, 12 per cent for business, five per cent for legal advice and remaining three pei cent for political and administrative purposes. Distance and purpose of interaction help in identifying the order of purposes. Results reveal that interactions for legal advice, political and administrative, business and social are the first order purposes for which mostly long distance interactions are made. Interaction for the search of job, education, services and religious purposes are identified as second order purposes. Movements for such purposes are generally from short and medium distance zones. The purposes, like, entertainment, medical care, drinks, services, business, education, legal advice and social interactions, are grouped in the third order purposes. For these purposes most of the interaction is from short distance zone. It has been observed that some of the functions move across these functional limits. Due to the presence of sub-orders within these, functional mobility for various purposes does not correspond closely with the function of decay (Fig. 5.5).

Impact of Distance on the Purpose of Movement

As explained above the purpose of movement varies largely with the distance. However, this generalisation is not true for all the purposes discussed above. To have an idea of the degree of relationship between the distance and the purpose, correlation exercise has been worked out. Results reveal that of the total 10 purposes, as many as eight have negative correlation and remaining two have positive correlation with the distance. Among the purposes registering negative correlation with the distance, purpose for work place interaction recorded the maximum relationship, r being –0.87. It is followed by the interaction for education purposes, r being –0.70. Legal advice, entertainment and alcoholic drinks have significant negative correlation, the values being 0.65. Low level of negative correlation was recorded in cases of three purposes, viz., interaction for business (r = –0.28), social (r = –0.24) and religious purposes (r = –0.15). Contrary to the above, movement for political and administrative purposes registered significant positive correlation with the distance, the level of correlation being +0.80. Similarly interaction for the search of job also registered positive correlation, the r value being +0.39 with the distance.

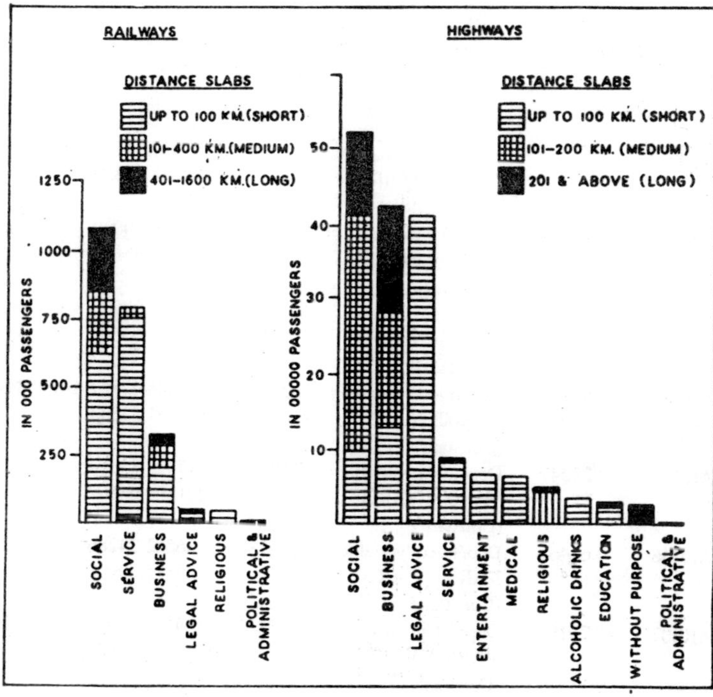

FIG. 5.5: Meerut city — Proportionate movement of population for various purposes 1984.

Having analysed the degree of association, cause and effect relationships were worked out to assess the impact of distance on various purposes. Results reveal (Table 5.15) that high degree of regression coefficients were registered for work place movement and political and administrative purposes, the respective regression coefficients being 0.75 and 0.64. Moderate regression coefficients were observed for five purposes viz., education, legal advice, entertainment, medical and alcoholic drinks; the regression coefficients being 0.49, 0.43, 0.42, 0.42 and 0.42, respectively. Low level of regression coefficients were recorded for four purposes, viz., search of job (r = 0.15), business (r = 0.079), social purposes (r = 0.058) and religious purposes (r = 0.023). Per unit change in the distance causes corresponding negative change (in lakh units) by 5.8 units in legal

TABLE 5.15: Correlation of Distance and Purpose of Movement

(A) Coefficient of Co-relation		*(B) Regression Coefficient*			
VAR001			R^2	*B*	*Beta*
VAR001	1.00000	Dependent variable = VAR002 VAR001 (Constant)	0.43190	–587329.71429 2749037.33333	–0.65719
VAR002	–0.65719	Dependent variable = VAR003 VAR001 (Constant)	0.42857	–94846.00000 442614.66667	–0.65465
VAR003	–0.65465	Dependent variable = VAR004 VAR001 (Constant)	0.42857	–94846.00000 442614.66667	–0.65465
VAR004	–0.65465	Dependent variable = VAR005 VAR001 (Constant)	0.07979	–109248.62857 1145651.53333	–0.28247
VAR005	–0.28247	Dependent variable = VAR006 VAR001 (Constant)	0.42857	–47422.85714 221306.66667	–0.65465
VAR006	–0.65465	Dependent variable = VAR007 VAR001 (Constant)	0.49237	–36852.34286 180515.20000	–0.70169
VAR007	–0.70169	Dependent variable = VAR008 VAR001 (Constant)	0.05827	–139105.40000 1547894.06667	–0.24139
VAR008	–0.24139	Dependent variable = VAR009 VAR001 (Constant)	0.74926	–182617.94286 917626.13333	–0.86560
VAR009	–0.86560	Dependent variable = VAR010 VAR001 (Constant)	0.02319	–14895.22857 140514.13333	–0.15227
VAR010	–0.15227	Dependent variable = VAR011 VAR001 (Constant)	0.15429	22523.57143 –35036.66667	0.39279
VAR011	0.39279	Dependent variable = VAR012 VAR001 (Constant)	0.64367	2530.57143 –5074.66667	0.80229
VAR012	0.80229				

VAR001 = Distance
VAR002 = Legal advice
VAR003 = Entertainment
VAR004 = Medical aid
VAR005 = Business
VAR006 = Alcoholic drinks
VAR007 = Education
VAR008 = Social
VAR009 = Service
VAR010 = Religious
VAR011 = Search for job
VAR012 = Political and administrative.

advice, 1.8 units in services, 1.4 units in social interactions and by 1 unit in business. Relatively low level of negative change, per unit of distance, was recorded (in thousand units) in cases of entertainment (94), medical (92), alcoholic drinks (47), education (36) and religious purposes (14). Contrary to the above, positive influences per unit of distance were recorded in the movements for the search of jobs (22.5) and for political and administrative purposes (2.5).

Distance and Modes of Conveyance

This exercise deals with the comparative assessment of the various kinds of modes for human interaction. Modes of the conveyance were used as dependent variables and distance as independent variable for this exercise. The exercise includes seven independent and nine dependent variables. Preferences for various modes of conveyance over different distance slabs are analysed in Table 5.16.

Modes for Medium and Long Distances

Modes used for travelling over medium to long distances are buses (roadways) and trains. In terms of human mobility, buses carry proportionately larger share of interaction than trains. It is evident from the data that there are two peaks of human movement through buses, viz., distance slabs 21 to 50 km and 101 to 200 km. The maximum frequency of traffic was recorded in the distance zone of 21 to 50 km. A bi-modal frequency curve marks the traffic movement through buses. Preference for bus, as a mode of conveyance, becomes negligible beyond a distance of 10 km. Traffic mobility, through buses, increases gradually up to 20 km becomes high between 21 to 200 km and decreases sharply beyond 200 km. This kind of traffic distribution is partly due to transport laws relating to stop-overs at certain intervals and partly due to discomfort of long journey through buses.

Trains are the most important modes used for long movements. Train services are available only on Meerut-Delhi, Meerut-Hapur and Meerut-Muzaffarnagar directions. Interaction through trains is very limited, i.e., 4.47 per cent of the total. However, most of the long distance movements are made through trains only. Due to low amount of travel towards long distances, overall movement through trains remains low. For the movement through trains, normal distribution curve having maximum frequency in the distance zone of 51 to 100 km

TABLE 5.16: Distance and Mode of Movement

Distance slabs (in km)	Percentage from the total of respective modes										Slab-wise share of movement
	Modes-wise interaction										
	Roadways (Bus)	*Private (Bus)*	*Matador*	*Train*	*Bicycle*	*Car, Motor-cycle, Scooter*	*Tonga*	*Tractor*	*Buggi*	*Others*	
Up to 10	0.57	6.10	58.62	1.25	68.26	71.34	90.85	48.55	85.13	100.00	23.00
11 - 20	5.61	10.93	15.35	10.04	21.54	20.28	6.28	35.38	10.32	—	10.63
21 - 50	32.40	30.60	26.03	20.25	10.20	8.38	2.87	16.07	4.55	—	27.09
51 - 100	14.25	52.37	—	38.52	—	—	—	—	—	—	22.51
101 - 200	30.46	—	—	9.72	—	—	—	—	—	—	10.40
201 - 500	16.40	—	—	14.50	—	—	—	—	—	—	6.02
More than 500	0.31	—	—	5.72	—	—	—	—	—	—	0.35
Total percentage	100.00	100.00	100.00	100.00	100.00	100.00	100.00	100.00	100.00	100.00	100.00
Total inter-action (Actual)	1,7208471	1,62,00,000	1,08,00,000	23,53,198	21,60,000	18,00,000	16,20,000	72,000	36,000	3,60,000	5,26,09,669
Percentage share to total inter-action	32.71	30.79	20.53	4.47	4.11	3.42	3.08	0.14	0.07	0.68	100.00

was recorded. Amount of traffic decreases sharply between 101 to 200 km. Traffic trend, however, further rises once again in the distance slab of 201 to 500 km. Thus, critical points of traffic increase are 100 and 300 km. Amount of traffic decreases considerably beyond these critical limits.

Modes for Short and Medium Distances

Private buses, mini-buses, matadors and taxis are the common modes of conveyance used for short and medium distance zones. These modes have number of halts and stops over short distance routes. Passengers mostly from nearby rural areas prefer to travel through these modes. Besides passenger traffic, some amount of light goods movement is also carried out through these modes. About 31 per cent of the total traffic movement is through private buses. Matadors and taxis accounted for 21 per cent of the total human movement. There is an apparent complementarity among the modes so far as movement of traffic is concerned. While private buses carry most of their traffic load (nearly 83%) to a range of 20 to 100 km, matadors and taxis carry most of their traffic load (about 74%) up to a range of 20 km. The function of distance decay operates in case of matadors and taxis whereas it does not operate in case of buses. Movement of passengers through private buses registered a regular rise up to distance of 100 km. As against this, movement through taxis and matadors fell down significantly beyond 10 km (Fig. 5.6).

Modes for Short Distances

Private cars, scooters, motorcycles, tractor, *tonga, buggi* and bicycles are the common modes of conveyance for short distances. These modes accounted for about 11 per cent of the total human interaction. Bicycles ranked first among these modes. It is followed by the private auto vehicles. *Tonga* accounted for about three per cent of the total interaction. Tractor and *buggi* are the common modes of transport for inter rural interactions. They are used as goods carrier for agricultural produce and also accounted for limited human interactions between rural areas and the city. These modes operate successfully within a range of 20 km from the city. These modes correspond closely with the function of distance decay.

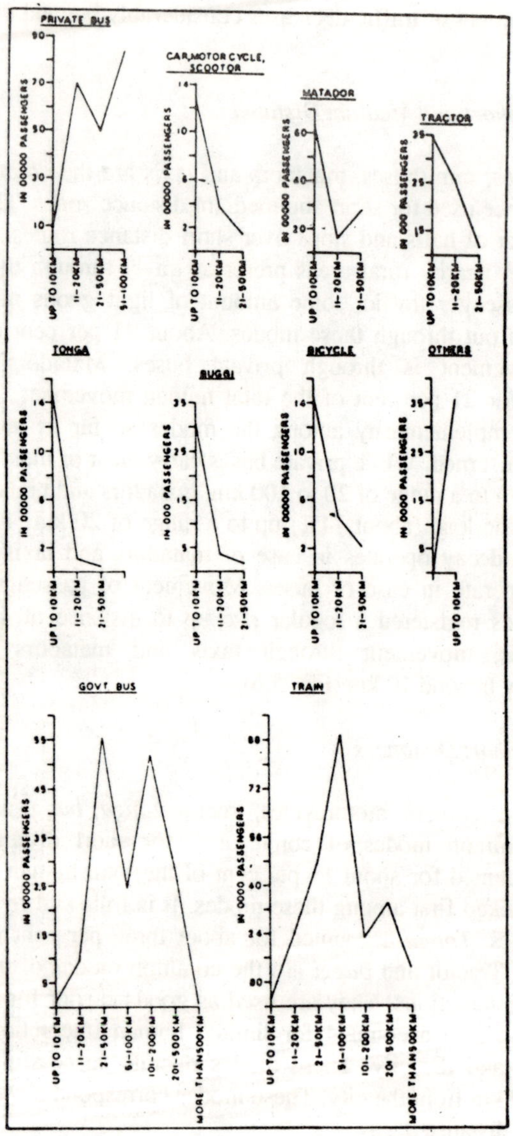

FIG. 5.6: Meerut city – Mode of movement and distance decay – 1984.

Overall Human Movement and the Distance Zones

Of the total interaction about one-third relates to a distance of 20 km from the city, about 50 per cent to a range of 21 to 100 km and remaining nearly 17 per cent beyond the distance of 100 km. It is, thus, evident that most of the human interaction is limited to a distance of 100 km. The relative importance of the modes over three distance zones reveals that matadors and taxis are the most preferred modes for the travel up to a distance of 20 km. Private and roadways buses are the most liked modes for the travel up to 100 km and trains for long journeys (beyond 100 km). It is, therefore, obvious from the study that alternate modes of transport have developed in Meerut region and have helped in increasing the intensity of interaction. Modes of conveyance are largely complementary and adhere to the function of distance decay in general. However, movement through private buses present an exception to this situation. The exceptional situation arises due to the inadequate number of bus services that causes added pressure of passengers through these buses. Most of these buses move overloaded i.e., much more than the recommended passenger load.

Impact of Distance on the Modes of Conveyance

Modes of conveyance play a significant role in bridging the gap of physical distance between the places and the regions. Different kinds of modes are used to approach different distance limits. Coefficient of correlations have been worked out to estimate the extent of relationship between the two variables. Results reveal that modes of movement, generally, have a negative correlation with the distance. However, the degree of correlation coefficient varies significantly. It ranges from –0.15 to –0.66. Tractor, matador, cars and bicycles have high negative correlation i.e., more than –0.52. In cases of private buses, *buggi* and *tonga*, the value of correlation coefficient ranged between –0.41 to –0.48. The low level of negative correlations (–0.15 to –0.37) were recorded in cases of trains, roadways buses and other modes of transport. The correlation values were found significant at all levels in cases of the first two categories whereas insignificant in case of the third one. Results of the regression coefficient reveal that it was 0.44 for tractor, 0.36 for matador, 0.297 for bicycles, 0.274 for car and 0.237 for buses. Regression coefficient for rest of the modes was less than 0.2. Per unit increase in the distance adversely affected the bus passengers by 6,917, matadors by 6,284, roadways buses by 1,538, bicycles by 1,296, car, scooter, motorcycles, etc., by 1,085, *tonga* by 995, train by 226, other modes by 223, tractor by 42 and *buggi* by 21 units. The above results and observations explain that high regression coefficients and high correlation values were recorded in cases of private modes of transport like tractor, bicycle, car, etc., whereas public modes of transport have low level of negative correlation and regression coefficients. Despite low correlations, the impact of distance influencing the number of units in each mode of transport remains high for public modes of transport than the private modes. The function of distance decay applies more closely with the private modes of transport compared to the public modes of transport. As such impact of distance over public modes of transport remains high despite low level of inverse relationship with the distance. Generally, modes of conveyance compete over short and medium distances whereas they monopolise over the long distances (Table 5.17).

Conclusion

The foregoing facts reveal that both the flow of commodities and the human movements are influenced by the factor of distance. Flow of commodities have, however, much longer space relations compared to the human movements. Industrial goods move over long distances and correspond positively with the price and the demand. On the contrary, agricultural goods move to relatively short distances and are inversely proportional to the distance and price and *vice versa* to the demand and the production. Most of the human movement is of daily frequency nature. It is mostly for the short distance zone and is for the purposes, like, legal advice, work place, social and trade purposes. Competitive and alternate modes of conveyance have developed to meet the growing demand of human interaction in this zone. Long distance movements are mostly for social, political and administrative purposes. Economic and educational interactions continue to remain in the second order of importance. Distance is, thus, proportional to the quality of the product and the order of the purpose.

TABLE 5.17: Correlation of the Distance and the Modes of Movement

(A) Coefficient of correlation	*(B) Regression coefficient*			
VAR001		R^2	*B*	*Beta*
VAR001 1.00000	Dependent variable =	0.02379	–1,538.24189	–0.15423
	VAR002 VAR001 (Constant)		27,72,593.84418	
VAR002–0.15423	Dependent variable =	0.23741	–6,917.92287	–0.48724
	VAR003 VAR001 (Constant)		37,27,518.52845	
VAR003–0.48724	Dependent variable =	0.36323	–6,284.10230	–0.60268
	VAR004 VAR001 (Constant)		28,26,609.46937	
VAR004–0.60268	Dependent variable =	0.03200	–226.24904	–0.17888
	VAR005 VAR001 (Constant)		3,82,390.58921	
VAR005–0.17888	Dependent variable =	0.29706	–1296.19891	–0.54503
	VAR006 VAR001 (Constant)		5,73,366.34792	
VAR006–0.54503	Dependent variable =	0.27491	–1085.07112	–0.52432
	VAR007 VAR001 (Constant)		4,78,807.38512	
VAR007–0.52432	Dependent variable =	0.17207	–995.19519	–0.41482
	VAR008 VAR001 (Constant)		4,34,732.73085	
VAR008–0.41482	Dependent variable =	0.44467	–42.34844	–0.66683
	VAR009 VAR001 (Constant)		18,936.89579	
VAR009–0.66683	Dependent variable =	0.19662	–21.98811	–0.44341
	VAR010 VAR001 (Constant)		9,634.71408	
VAR010–0.44341	Dependent variable =	0.14053	–223.19475	–0.37487
	VAR011 VAR001 (Constant)		97,024.07002	
VAR011–0.37487				

VAR001 = Distance
VAR002 = Bus (Roadways)
VAR003 = Bus (Private permit)
VAR004 = Matador
VAR005 = Train
VAR006 = Bicycle
VAR007 = Car/Scooter/Motorcycle
VAR008 = *Tonga*
VAR009 = Tractor
VAR010 = *Buggi*
VAR011 = Others.

6

DETERMINANTS OF INTERACTION

INTRODUCTION

THIS CHAPTER deals with the factors determining the commodity flows and the human movements. Apparently both, the interactions goods and human movements, are determined by the factors of transport, communication, technology and institutions. Availability of the adequate and efficient modes of transport and communications is the pre-requisite of interactions. Extent of interaction depends on the potentials of a place which rests on the level of accessibility, extent of industrial as well as agricultural development and the institutional concentration. As institutions offer services, aids, knowledge and information a sizable proportion of the interaction is generated due to their concentration. Institutions act as catalysts for socio-economic modernisation in an area.

An attempt has been made here to measure the extent of influence exerted by the group of determinants separately as well as jointly. Determinants are considered to have three main dimensions, viz., technology, transport and communications and institutions. The area of research is district Meerut whereas unit of the study is block. Data used in this exercise refer to the periods of 1979 and 1984. In all 30 indicators have been selected for this purpose. Indicators representing group of technological determinants are six, transport and communication determinants are eight and institutional determinants are 16. By using the scale of unit area or unit population, values of the indicators were calculated. Since the derived values of the indicators were large, precise comparison and addition was not possible. It was, therefore, necessary to reduce the values to their reasonable size. Values of the indicators were divided by their respective means to obtain the smaller size. Following the above procedure, indicators were normalised. Based on the perceptible gaps in the array of the values, units were classed as very high, high, medium, low and very low. The correlation coefficients have been calculated to measure the influence of the groups separately whereas composite scores have been calculated to measure the influence of groups collectively. The groups of determinants and indicators are enlisted below:

Technological Determinants

(i) Number of tractors per ten thousand hectares of net sown area,
(ii) Number of tube-wells per ten thousand hectares of net sown area,
(iii) Number of pumpsets per ten thousand hectares of net sown area,
(iv) Distribution of 100 tonnes of fertilizers per ten thousand hectares of net sown area,
(v) Number of animals per thousand units of reported area, and
(vi) Value of industrial production per worker.

Transport and Communication Determinants

(i) Length of metalled road per 1,000 sq. km of area,
(ii) Length of road per lakh of population,
(iii) Number of post offices per lakh of population,
(iv) Number of railway halts per lakh of population,
(v) Number of bus stops per lakh of population,
(vi) Number of telegraph offices per lakh of population,
(vii) Number of telephone offices per lakh of population, and
(viii) Number of public call offices per lakh of population.

Institutional Determinants

(i) Number of seed godowns per 10 villages,
(ii) Average agricultural land (in hectares) per village,
(iii) Average number of *gobar* gas plants per village,

(iv) Utilization of fertilizers per hectare of net sown area,
(v) Number of animal hospitals per lakh of animal population,
(vi) Number of animal development centres per 10 villages,
(vii) Number of artificial semen centres per lakh of animal population,
(viii) Number of piggery units per 10 villages,
(ix) Number of milk societies per 10 villages,
(x) Distribution of fishery units per village,
(xi) Number of agricultural loan societies per 10 villages,
(xii) Amount of deposit per capita of population,
(xiii) Distribution of development loan per capita of population,
(xiv) Number of family planning centres per 10 villages,
(xv) Percentage of literate population to total population, and
(xvi) Number of fair price shops per 10 villages.

Detailed analysis of each determinant group is given below:

Technological Determinants

As process of development revolves round the technology, it forms the pivotal base for all kinds of progress. Level of technological achievement determines the nature of interaction in a region. Results reveal that technological determinants (Table 6.1) ranged between the score values 0.59 in case of block Hastinapur to 2.76 in case of block Meerut during 1979. While blocks of Meerut, Kharkhauda, Baghpat and Mawana registered very high scores, blocks of Machhra, Rajpura and Chhaprauli registered the high scores of technological determinants. The moderate technological determinants were recorded in the blocks of Baraut, Jani, Rohta and Daurala. The blocks of Saroorpur, Pilana and Khekra recorded in low scores of technological determinants whereas blocks of Hastinapur, Parikshhatgarh, Sardhana and Binoli recorded the very low scores of technological determinants.

Over a period of five years, i.e., in 1984, significant changes were recorded in the scores of technological determinants. The score values ranged between 0.58 (in block Hastinapur) to 1.95 (in block Daurala). Blocks of Daurala, Chhaprauli, Meerut and Mawana recorded very high scores and formed an areal cluster along Meerut-Mawana axis. Similarly blocks of Baghpat and Baraut formed a cluster of high determinants towards the west whereas block Machhra formed an isolated pocket towards the east. Moderate technological achievements

TABLE 6.1: District Meerut — Determinants of Interaction — 1979

Block	Determinants											
	Technological			*Transport*			*Institutional*			*Overall determinants*		
	Value	*Rank*	*Class*	*Value*	*Rank*	*Class*	*Value*	*Rank*	*Class*	*Value*	*Rank*	*Class*
1. Meerut	2.76	1	VH	3.09	1	VH	0.97	9	M	6.82	1	VH
2. Jani	0.92	8.5	M	0.96	7	H	0.88	11	M	2.76	11	M
3. Rohta	0.85	10	M	0.77	12.5	L	0.86	12	L	2.48	13	L
4. Rajpura	1.01	5	H	0.66	17	VL	0.69	17	VL	2.35	14	L
5. Kharkhauda	1.44	2	VH	0.91	8	M	1.00	8	M	3.35	4	VH
6. Mawana	1.03	4	VH	1.05	4	VH	1.31	2	VH	3.39	3	VH
7. Hastinapur	0.59	18	VL	0.98	6	H	0.67	18	VL	2.24	15	VL
8. Parikshhatgarh	0.75	15	VL	0.70	15	VL	0.74	16	VL	2.19	18	VL
9. Machhra	1.00	6	H	0.83	11	M	1.14	5	H	2.98	7	H
10. Sardhana	0.72	16	VL	1.14	3	VH	1.23	3.5	VH	3.09	6	H
11. Saroorpur	0.78	14	L	0.90	9	M	1.23	3.5	VH	2.91	9	M
12. Binoli	0.66	17	VL	0.69	16	VL	0.85	13	L	2.20	16.5	VL
13. Daurala	0.83	11.5	M	1.23	2	VH	0.91	10	M	2.97	8	M
14. Baghpat	1.10	3	VII	1.04	5	H	1.03	7	H	3.17	5	H
15. Baraut	0.92	8.5	M	0.77	12.5	L	1.10	6	H	2.79	10	M
16. Chhaprauli	0.99	7	H	0.75	14	L	1.83	1	VH	3.57	2	VH
17. Khekra	0.83	11.5	L	0.87	10	M	0.80	14	L	2.50	12	L
18. Pilana	0.80	13	L	0.63	18	VL	0.77	15	VL	2.20	16.5	VL

Note: Further details about the variables see Appendices 3A, 3B and 3C.

TABLE 6.2: District Meerut — Determinants of Interaction — 1984

Blocks	Determinants											
	Technological			*Transport*			*Institutional*			*Overall determinants*		
	Value	*Rank*	*Class*	*Value*	*Rank*	*Class*	*Value*	*Rank*	*Class*	*Value*	*Rank*	*Class*
1. Meerut	1.27	3	VH	2.69	1	VH	1.04	7	H	5.00	1	VH
2. Jani	0.96	8	M	0.93	10.5	M	0.92	11	M	2.81	10	M
3. Rohta	0.87	13	L	0.94	8.5	M	0.93	10	M	2.74	12	L
4. Rajpura	0.95	9.5	M	0.78	14	L	1.06	5	H	2.79	11	M
5. Kharkhauda	0.89	12	L	0.60	15	VL	1.00	9	M	2.49	14	L
6. Mawana	1.20	4	VH	1.09	5	H	1.22	3	VH	3.51	3	VH
7. Hastinapur	0.58	18	VL	0.96	7	H	0.80	17	VL	2.34	16	VL
8. Parikshhatgarh	0.78	15	VL	0.79	13	L	0.86	13	L	2.43	15	VL
9. Machhra	1.02	6.5	H	0.80	12	L	1.12	4	VH	2.94	8	M
10. Sardhana	0.95	9.5	M	1.12	3.5	VH	1.04	7	H	3.11	6	H
11. Saroorpur	0.83	14	L	1.12	3.5	VH	1.26	2	VH	3.21	5	H
12. Binoli	0.71	17	VL	0.57	16	VL	0.87	12	L	2.15	18	VL
13. Daurala	1.95	1	VH	1.58	2	VH	0.85	14	L	4.38	2	VH
14. Baghpat	1.02	6.5	H	1.05	6	H	0.84	15.5	VL	2.91	9	M
15. Baraut	1.03	5	H	0.93	10.5	M	1.04	7	H	3.00	7	H
16. Chhaprauli	1.34	2	VH	0.53	18	VL	1.50	1	VH	3.37	4	VH
17. Khekra	0.90	11	M	0.94	8.5	M	0.79	18	VL	2.63	13	L
18. Pilana	0.76	16	VL	0.56	17	VL	0.84	15.5	VL	2.16	17	VL

Note: Further details about the variables see Appendices-3D, 3E and 3F.

were observed in the blocks of Khekra, Jani and Sardhana. While blocks of Kharkhauda, Saroorpur and Rohta registered the low scores, blocks of Hastinapur, Parikshhatgarh, Binoli and Pilana registered the very low scores of technological determinants (Table 6.2).

Over two points of time technological determinants registered the following changes:

(i) Eleven out of 18 blocks in the district registered increase in their score values during 1979-84. Rest of the seven blocks, namely, Meerut, Rajpura, Kharkhauda, Hastinapur, Baraut, Chhaprauli and Pilana, registered fall in their index values. Such a change explains an overall improvement in the situation in terms of technological determinants. However, rapid increase in the population, due to growing urban industrial development, caused corresponding fall in the index values of some of the blocks,

(ii) While blocks of Mawana, Hastinapur, Saroorpur and Binoli maintained the same rank position in 1984 which they had in 1979, blocks of Jani, Sardhana, Daurala, Baraut, Chhaprauli and Khekra registered rise in their ranks. Rest of the blocks (seven) registered fall in their earlier ranks, and

(iii) In terms of changes in the levels of technological determinants, it was noted that as many as eight blocks did not register any change in their earlier levels. However, rest of the 10 blocks registered change from their earlier positions. While blocks of Khekra, Chhaprauli, Baraut, Daurala and Sardhana registered upward shift; blocks of Rohta, Rajpura, Kharkhauda, Baghpat and Pilana registered downward shift. It was, further, noted that shift towards either side (positive or negative level) was by one order in as many as seven out of 10 cases. Major shifts were recorded only in cases of three blocks. While blocks of Daurala and Sardhana moved up by two orders, block of Kharkhauda slipped down by the three orders (Fig. 6.1).

Transport and Communication Determinants

Transport and communications provide the means for interaction. Development is an interdependent phenomenon in which linkages (regional as well as inter-regional) form the basis for inter-connections between the places at reduced cost and time. Connectivity, linkages

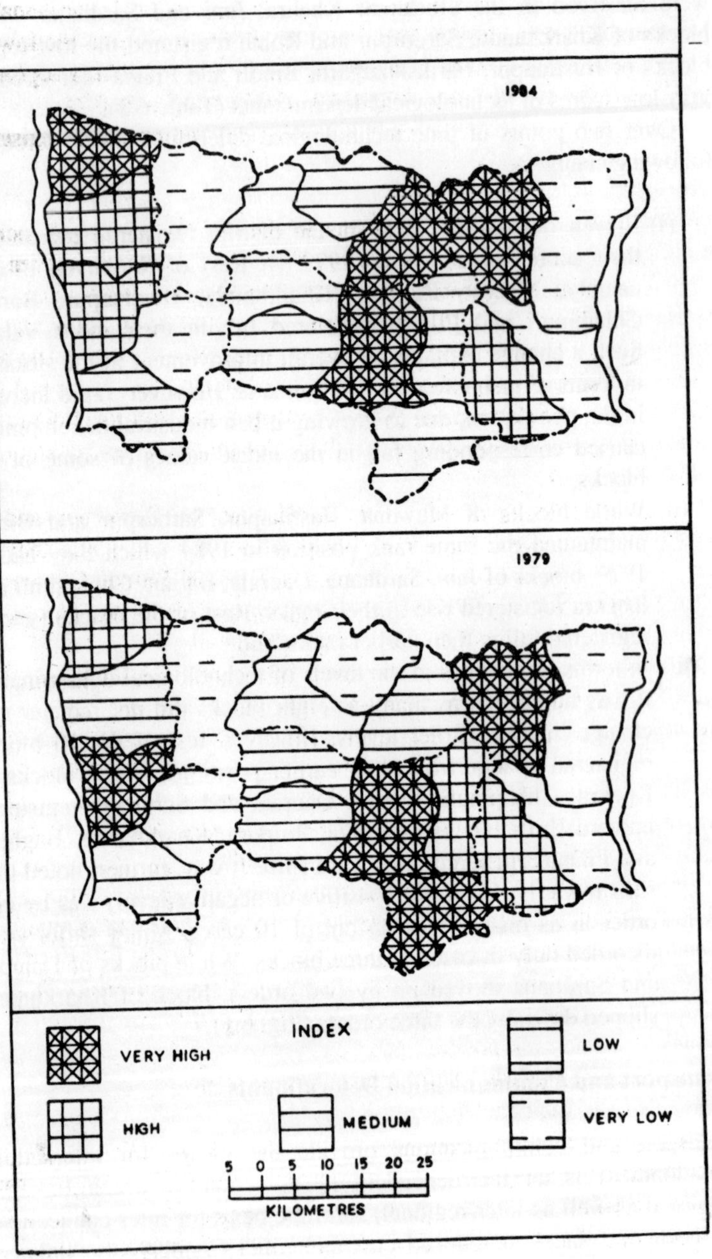

FIG. 6.1: Technological determinants in Meerut.

and communication facilities are, therefore, proportional to the index of regional development. As such, factors of transport and communication act as an important determinant in explaining commodity flows and human interactions. Scores for this group of indicators ranged between 0.63 (in block Pilana) to 3.09 (in block Meerut) during 1979. The values of the transport and communication determinants are generally higher compared to the values of technological determinants for 1979. Blocks of Meerut, Daurala, Sardhana and Mawana registered very high; blocks of Baghpat, Jani and Hastinapur high and the blocks of Saroorpur, Machhra, Kharkhauda and Khekra medium scores in this group of determinants. The low scores of determinants were registered in the blocks of Baraut, Chhaprauli and Rohta whereas very low scores of determinants were recorded in the blocks of Binoli, Pilana, Parikshhatgarh and Rajpura (Table 6.1).

The situation over a period of five years registered minor changes. Scores for 1984 ranged from 0.53 (in block Chhaprauli) to 2.69 (in block Meerut). Thus, there was an overall decline in the score values over two points of time. Relative sectoral position of the transport and communication determinants have, however, remained unchanged. Blocks of Meerut, Sardhana, Saroorpur and Daurala recorded very high score values; blocks of Mawana, Hastinapur and Baghpat high score values and blocks of Jani, Rohta, Baraut and Khekra medium score values. The low score values were recorded in Rajpura, Parikshhatgarh and Machhra blocks and that of very low index values were recorded in the blocks of Pilana, Chhaprauli, Binoli and Kharkhauda (Table 6.2).

Temporal changes registered an overall decline in the composite scores of the determinant groups. However, blocks of Rohta, Rajpura, Mawana, Parikshhatgarh, Saroorpur, Daurala, Baghpat, Baraut and Khekra registered rise in their scores of 1984 compared to 1979. In terms of their relative rank, blocks of Meerut, Binoli and Daurala recorded the same rank position in 1984. Seven blocks namely, Rohta, Rajpura, Parikshhatgarh, Saroorpur, Baraut, Khekra and Pilana, registered rise in their ranks during 1979-84. Remaining blocks (eight) slipped down from their previous rank positions. Changes brought in the levels of areal units suggest a balanced growth of this sector. As many as eight blocks registered the same levels of determinants over the two points of time. Five blocks, viz., Rohta, Rajpura, Parikshhatgarh, Saroorpur and Baraut, recorded the positive shift in the

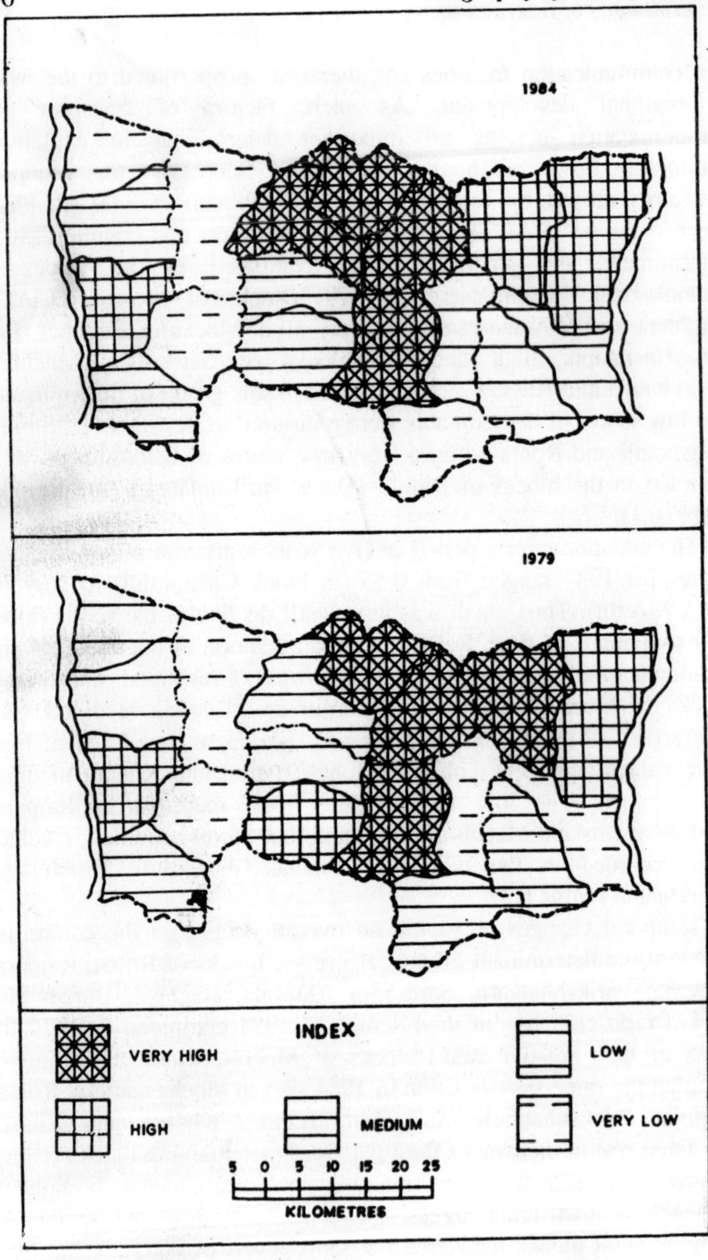

FIG. 6.2: Transport determinants in Meerut.

levels of their determinants. Thus, most of the blocks have either maintained the same level of determinant or registered upward shift in the hierarchic order (Fig. 6.2).

Institutional Determinants

Institutions act as centres of information flows, capital flows, services and amenities. Institutions, thus, act as one of the determinants of interactions. Institutional concentration leads to the increased interaction whereas even distribution leads to the normal scale of interaction among the units. While technological determinants introduce the element of change in the process of production, transport and communication determinants accelerate the links between surplus and deficit areas and widen the scope of interaction. Similarly, institutional determinants help in creating the basic conditions within which interaction may take place. Institutional determinants, thus, channellise the services, information and amenities among the places. An account of the institutional determinants influencing the development is given in Table 6.1. Results reveal that institutional determinants have more regular distribution compared to the first two groups of the determinants. The score values ranged from 0.67 in block Hastinapur to 1.83 in block Chhaprauli. Blocks of Chhaprauli, Mawana, Sardhana and Saroorpur registered very high scores. Next in order of importance were the blocks of Baghpat and Machhra (values from 1.03 to 1.14). Levels of medium score values were recorded in the blocks of Meerut, Jani, Kharkhauda and Daurala. While low scores of institutional determinants were recorded in the blocks of Rohta, Binoli and Khekra; very low scores of determinants were recorded in the blocks of Pilana, Rajpura, Parikshhatgarh and Hastinapur. Thus, north-western part of the district registered high scores of institutional determinants whereas the eastern part registered very low scores of the determinants. Central part of the district recorded average scale of institutional determinants.

Table 6.2 reveals that index values registered nominal changes over a period of five years. Determinant scores ranged between 0.79 in block Khekra and 1.50 in block Chhaprauli during 1984. Thus, lower limit rose up to 0.79 (1984) from 0.67 (1979) whereas upper limit slipped down to 1.50 (1984) from 1.83 (1979). Blocks of Chhaprauli, Saroorpur, Machhra and Mawana observed very high scores whereas Meerut, Rajpura, Sardhana and Baraut observed high scores of determinants. Blocks of Jani, Rohta and Kharkhauda recorded the medium level of determinants. While low scores were recorded in three scattered pockets of Binoli, Daurala and Parikshhatgarh; very low scores were recorded in the south-western and the north-eastern areas of the district. The following changes have been recorded during 1979-84:

(i) In terms of changes in the scores, nine blocks recorded rise and eight fall, while the remaining one recorded no change. Among the nine blocks registering rise in score values, four blocks have formed a cluster near Meerut. Blocks along the Hindon and the Ganga rivers have also experienced rise in their scores,

(ii) Relative position of the ranks reveals that two blocks (of Chhaprauli and Jani) maintained the same rank whereas eight blocks recorded rise and the remaining eight fall over the two points of time. Since the extent of shift has been limited in case of the levels compared to the scores and the ranks, as many as nine out of 18 blocks registered no shift in their levels. Among the nine blocks registering shift, five blocks moved up and four blocks slipped down in their respective levels. Against this general trend, special trends were also recorded in the blocks Rajpura and Baghpat where former recorded and upward shift by three levels and the latter recorded the downward shift by the same levels (Fig.6.3).

Overall Determinants

With the help of composite scores of three groups of determinants, and exercise has been attempted to workout the overall scores of determinants for each unit. This exercise enables comparison between the units and determinant groups over two points of time. The results reveal that there are significant variations in the scores of the determinants. The score values ranged between 2.19 in block Parikshhatgarh to 6.82 in block Meerut. Twelve out of 18 blocks recorded the score point below three, five blocks between three and 3.57 and one, i.e., Meerut block, recorded an exceptionally high score point of 6.82. Apparently, there are marked variations in the scores of areal units and the groups of the determinants. Extent of variation reduces significantly (1.38) if scores of block Meerut are excluded from the group analysis. Exceptionally high scores of Meerut indicate

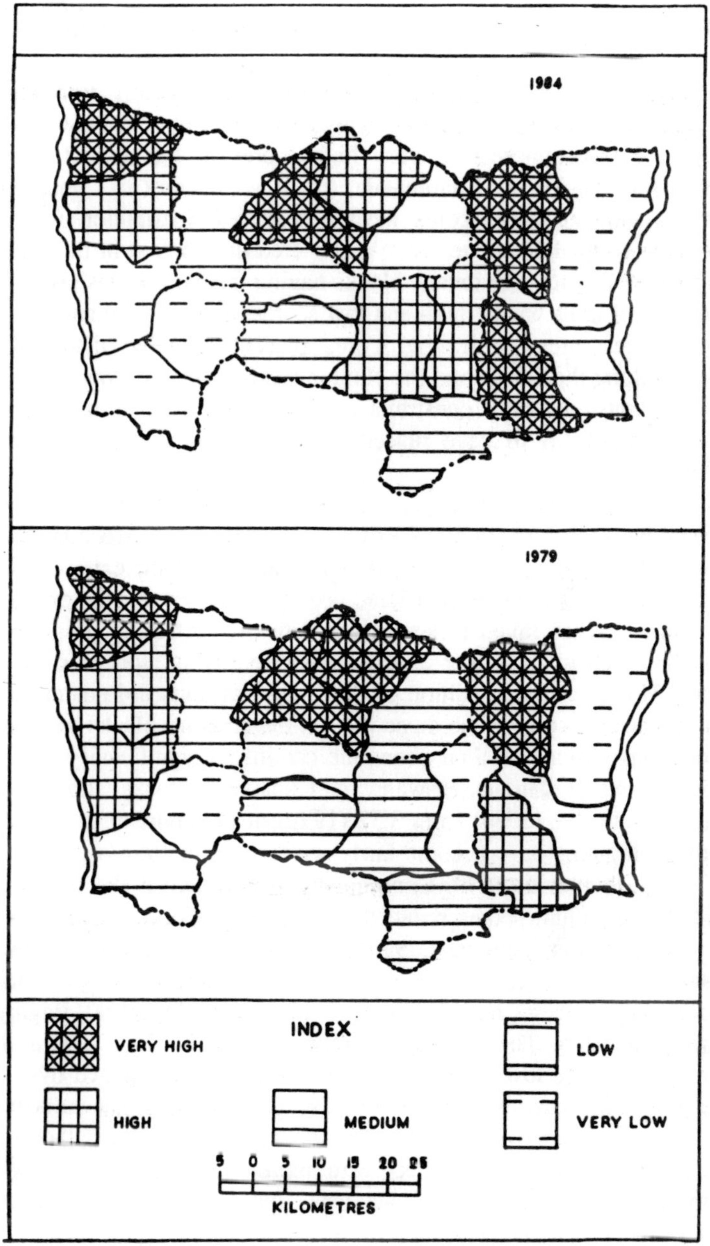

FIG. 6.3: Institutional determinants in Meerut.

concentration of growth impulses over here. Consequently, imbalances are caused in the process of the regional growth. Figure 6.4 shows that there are three pockets of very high scores in the district. These pockets are (i) Meerut-Kharkhauda, (ii) Mawana and (iii) Chhaprauli. Blocks showing high scores have a scattered distribution whereas blocks with medium scores have a clustered distribution in the central part of the district. Obviously, blocks having medium scores lay in the interface of units having high and low scores of the determinants. Like high scores, areas with low scores present scattered distribution. Blocks along the Ganga and the Hindon rivers have very low scores of determinants. Natural constraints, such as, foods, droughts and wastelands, seem to have placed these areas at the lowest scores (Table 6.1).

Over a period of five years (1984) the index of overall scores ranged between 2.15 (in block Binoli) to five (in block Meerut). There has been a decline in the scale of determinants. Both the extremes, i.e., maximum and minimum, registered decline. It shows that the extent of determinants has failed to keep pace with the increasing pressure of population. Results further explain that disparity is reduced over the years. However, the condition of the backward units did not improve due to the reduction in the disparity. On the contrary blocks with high scores have improved their economic conditions significantly. This is true in cases of Daurala, Mawana and Chhaprauli blocks whose upper limit rose to five (1984) from 3.37 (1979). It is important to note that growth impulses have become fairly widespread in 1984 compared to their distribution in 1979. Consequently gap between the maximum and the minimum scores reduced considerably. Further, blocks with high scores, like, Sardhana, Saroorpur and Baraut, presented lesser variations compared to the blocks having very high scores. Medium index values were recorded in the blocks of Machhra, Baghpat, Rajpura and Jani. The index values ranged between 2.79 to 2.94 in this group. Scores of low to very low index remained unchanged over the years. Block Kharkhauda presents exception to this generalisation. There has been an overall decline in the score values of these blocks. It is, therefore, clear that blocks with more than the average scores improved their conditions whereas those below the average deteriorated in their conditions over the years (Table 6.2).

The process of change over five years presented significant variations which led to the following observations:

(i) As many as 10 out of 18 blocks have registered a rise in their respective scores of 1984 over 1979. Of these, Daurala made remarkable progress by scoring 4.38 in 1984 as against 2.97 in 1979. Such an increase in the scores is due to the structural changes introduced in this area. With the establishment of Modi Rubber Complex and other industrial units, index of overall determinants registered steep rise. While most of the blocks have registered rise, a few of them have experienced fall in their respective scores. Blocks of Rajpura, Kharkhauda, Machhra and Meerut are characterised by high population pressure, low to average industrial development and average to high agricultural development. As such, despite increase in absolute growth impulses, scores could not improve in these blocks. Because of the increasing pressure of population, growing menace of salinity and wasteland, Binoli and Pilana blocks registered a fall in their score values. Similarly, because of the poor functional and communicational links, Chhaprauli and Baghpat blocks registered a fall in their score values. Thus, decline in the scores of the overall determinants have been due to the increasing pressure of population near Meerut, flood and wastelands along main rivers and inadequate functional and communicational links in intervening areas,

(ii) In terms of relative rank positions blocks of Meerut, Mawana and Sardhana maintained their rank positions; blocks of Jani, Rohta, Parikshhatgarh, Saroorpur, Daurala and Baraut registered upward shift in their ranks and remaining blocks registered a fall in their ranks over two points of time (1979 and 1984), and

(iii) It is important to note that most of the blocks registered the same level of determinants, four blocks, namely, Rajpura, Saroorpur, Daurala and Baraut registered upward rise in their hierarchic order. The maximum upward shift was in the case of Daurala which recorded 8th rank and medium level in 1979 and second rank and very high level in 1984. On the contrary, blocks of Kharkhauda, Machhra and Baghpat slipped down in their hierarchic level over two points of time (Fig. 6.4).

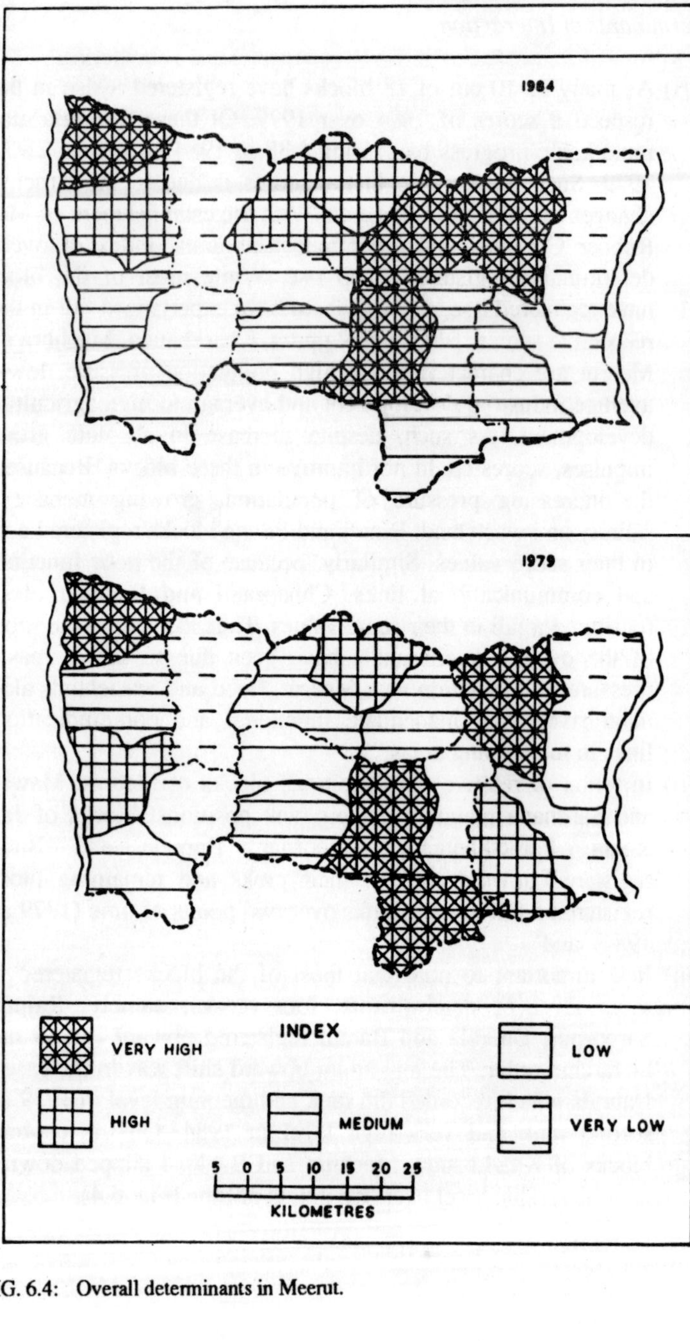

FIG. 6.4: Overall determinants in Meerut.

Nature of Imbalances in the Determinants of Interaction

It is now accepted that balanced sectoral growth is essential for removing regional as well as economic disparity. The foregoing discussion makes it clear that determinants vary in sectoral as well as in regional terms. Sectoral growth has not been uniform over all the blocks because of the varying socio-economic and geographical conditions. This section is, therefore, devoted to the study of imbalances in the district. Following assumptions have been made to workout the nature of imbalances in the district (Table 6.3):

(i) Blocks registering almost similar performance in institutional as well as non-institutional determinants have been termed as areas experiencing the balanced growth whereas varying performances in the determinants have been termed as areas experiencing imbalanced growth,

(ii) For the convenience of comparison and analysis areas with imbalanced growth have been re-grouped into areas of (a) relatively advanced in non-institutional determinants and (b) relatively advance in institutional determinants.

TABLE 6.3: Imbalances in the Determinants

Levels of development	*Sectoral growth*		
		Relatively advanced in	
	Balanced	*Non-Institutional sectors*	*Institutional sectors*
Very high	Meerut, Mawana, Chhaprauli	Daurala	—
High	Baraut	Sardhana, Saroorpur	—
Medium	Jani	Baghpat	Rajpura, Machhra
Low	Rohta	Khekra	Kharkhauda
Very low	Parikshhatgarh, Binoli, Pilana.	Hastinapur	—

Areas Experiencing Balanced Growth of the Determinants

About half of the blocks (9) in the district have registered balanced sectoral growth of the determinants. This type of growth is observed at various levels of determinants. For instance, Meerut, Mawana and Chhaprauli at very high level, Baraut at high, Jani at medium, Rohta at low and Parikshhatgarh, Binoli and Pilana at the very low level of determinants have observed this kind of sectoral growth. It is, thus, clear that areas with balanced growth of the determinants have three clusters, viz., Meerut, Baraut and Mawana. It may be generalised that most of the blocks registering balanced growth are either at very high level or at the very low level of determinants. The tendency towards balanced sectoral growth is observed either at the initial level when economy remains under-developed or at the higher level when economy attains accelerated growth due to economic diversification.

Areas with Relatively Advanced in Non-Institutional Determinants

About one-third of the blocks have registered relatively higher growth in non-institutional determinants. This type of sectoral growth was observed in the north-central blocks including Sardhana, Saroorpur and Daurala, in the south-western blocks of Baghpat and Khekra and in the north-eastern blocks of Hastinapur and Daurala. Of these, Daurala registered the level of very high determinants, Sardhana and Saroorpur – high, Baghpat – medium, Khekra – low and Hastinapur – very low. It is, thus, clear that growth of non-institutional sectors is associated more with high and low levels.

Areas with Relatively Advanced Institutional Determinants

This type of advancement has been observed in cases of three blocks, viz., Rajpura, Machhra and Kharkhauda. These blocks have an advantage of being near Meerut, the chief centre of socio-economic interaction. Among these blocks, Rajpura and Machhra are at the medium level whereas Kharkhauda is at the low level of non-institutional determinants. It is, thus, established that relatively advanced institutional growth is associated with the areas registering low to medium growth of determinants. Departure from the normal (positive as well negative) leads to the balanced growth or growth in non-institutional determinants. Institutional advancement is, therefore, essential for achieving balanced growth of the determinants. It is important to note that industrial development is concentrated along north-sough axis, agricultural development along the western blocks and institutional development towards south-eastern blocks of the district.

Disparities in the Determinants

Disparities are caused due to uneven growth of the determinants. There are varying responses to these determinants. It needs to be analysed:

(i) the precise degree of association between the disparity levels and the groups of the determinants,
(ii) the disparity influences over areal growth, and
(iii) the accentuation of the disparity over time.

It is in the light of spatial and sectoral advancements of the determinants that an exercise needs to be attempted to explain the nature of disparity. Detailed account of sectoral disparity is given below (Table 6.4).

TABLE 6.4: Disparities in Sectoral Determinants

Sector	*Levels of development*					
	Very high	*High*	*Medium*	*Low*	*Very low*	*Over all*
Technological	–0.98	–0.01	–0.03	+0.01	+0.04	–0.80
Transport	–0.47	+0.05	–0.07	0	0	–0.30
Institutional	+0.22	–0.09	–0.04	–0.04	–0.05	–0.45
Overall	–1.84	+0.02	–0.06	+0.10	+0.23	–1.78

Disparities in the Growth of Technological Determinants

Technological determinants registered decline in the disparities during 1979-84. Range of technological determinants reduced to 1.37 in 1984 from 2.17 in 1979. Regions with high technological determinants recorded the minimum level of disparity. Decline in the disparity is due to the corresponding rise in the minimum as well as maximum score points during 1979-84. Areas with medium level of technological determinants observed decline in the disparity trend. However, the extent of decline was lower than the first two groups of the determinants. On the contrary, disparities increased slightly in areas experiencing low level of technological determinants. Due to increase in actual scores during the period under study, gap between maximum and minimum limits increased as a natural consequence. The situation, in which minimum point falls down and maximum point rises up, explains increase in the disparity. Areas with low to very low level of technological determinants have also registered increase in the sectoral disparity, though reason for the increase is different to that of the first. While increase in the disparity has been due to slow growth in case of the former, it has been due to steep fall in the minimum and slow rise in the maximum limits in case of the latter (Table 6.5).

TABLE 6.5: Temporal Trends of Disparity

	Range of development					
	1979			1984		
Sectors	*Minimum*	*Maximum*	*Extent of variation*	*Minimum*	*Maximum*	*Extent of variation*
Technological	0.59	2.76	2.17	0.58	1.95	1.37
Transport	0.63	3.09	2.46	0.53	2.69	2.16
Institutional	0.67	1.83	1.16	0.79	1.50	0.71
Overall	2.19	6.82	4.63	2.15	5.00	2.85

Disparities in the Transport and Communication Determinants

Transport and communication determinants, alike technological, experienced decline in the extent of the disparity. Level of disparity reduced to 2.16 (1984) from 2.46 (1979). The growth of this sector is not in conformity with the population growth and dependency ratios. Sectoral performance presents variation in the disparity at all levels of sectoral growth. Decline in the disparity was about 0.47 in case of units registering very high scores, 0.05 in case of units with high scores and 0.07 in case of units with medium scores. On the contrary disparity remained unchanged in case of units registering low to very low scores. While increase in the disparity is attributed to the increase in the range of score limits, reduction in the disparity is partly due to the slow increase and partly due to growing under-development over the years. For instance, while minimum score limit remained unchanged (i.e., 1.09), maximum score limit increased slightly (i.e., 1.09, 1.04) in case of the latter.

Disparities in the Institutional Determinants

Disparities in the institutional determinants reduced by 0.45 score points during 1979-84. There has been a decline in the disparity at all levels of institutional determinants. For instance, disparity reduced by 0.22 score points in case of very high, 0.09 in case of high, 0.04 in case of medium, 0.04 in case of low and 0.05 in case of very low institutional determinants. Facts reveal that reduction in the disparity has been maximum in cases of very high determinants and low in cases of medium to low level of determinants. This is an important trend along which disparity reduces with the institutional determinants.

Disparities in the Overall Determinants

Two distinct disparity trends were observed in the overall determinants. Disparities declined by 1.78 score points during 1979-84. However, decline in the disparity did not remain the same at all levels. For instance, it decreased by 1.84 score points at very high level and 0.06 at medium level. On the contrary, disparity increased by 0.02 score points at the high level, by 0.10 at the low level and 0.23 at the very low level. Thus, overall disparity presents decline which is inversely proportional to the levels of the determinants. Disparity declined sharply between areas registering medium to very high scores of determinants. On the contrary, it registered nominal increase in the disparity in areas registering low to very low level of overall determinants (Table 6.6).

TABLE 6.6: Disparity in the Levels of Determinants

Levels of determinants	*Class disparity*		*Nature of temporal change*
	1979	*1984*	
Very high	3.47	1.63	Reduced
High	0.19	0.21	Increased
Medium	0.21	0.15	Reduced
Low	0.15	0.25	Increased
Very low	0.05	0.28	Increased

It is further observed that areas registering high scores have a tendency to grow more or less at the same rate over a period of time whereas areas with the low scores have a tendency to register unequal growth of the units. While Kharkhauda-Daurala axis registered high to very high disparity, Jani-Sardhana axis with inclusion of Pilana and Binoli blocks observed very low disparity. Blocks of Khekra and Chhaprauli on the west, Parikshhatgarh on the east and Saroorpur and Rohta on the centre registered medium disparity. As against this Mawana-Hastinapur axis registered low level of disparity (Table 6.7).

TABLE 6.7. Nature of Disparity in the Overall Determinants

Sl. No.	*Blocks*	*Nature of disparity*	*Sl. No.*	*Blocks*	*Nature of disparity*
1.	Meerut	–1.82	10.	Sardhana	+0.02
2.	Jani	+0.05	11.	Saroorpur	+0.23
3.	Rohta	+0.26	12.	Binoli	–0.05
4.	Rajpura	+0.44	13.	Daurala	+1.41
5.	Kharkhauda	–0.86	14.	Baghpat	–0.26
6.	Mawana	+0.12	15.	Baraut	+0.21
7.	Hastinapur	+0.10	16.	Chhaprauli	–0.20
8.	Parikshhatgarh	+0.24	17.	Khekra	+0.13
9.	Machhra	–0.04	18.	Pilana	–0.04

Determinants and their Correlates

An exercise of correlation coefficient has been worked out to analyse the extent of inter-relationship between the groups and the scores of overall determinants. Separate exercises were attempted to calculate the values of correlation coefficients for the periods 1979 and 1984. The temporal trends explain the disparity behaviour over the two points of time. Overall determinants recorded positive association with all the groups of the determinants, though degree of response varied significantly from one to the other. While technological determinants recorded the maximum correlation of 0.94, institutional determinants recorded the minimum correlation of 0.32 with the overall determinants during 1979. Alike technological determinants, transport and communication determinants have also recorded significantly high positive correlation (0.93) during 1979. In case of inter-correlations between the groups, technological groups of determinants recorded high positive correlation with the transport group, r value being 0.87. On the contrary, institutional groups of determinants recorded insignificant positive correlation with the remaining two groups. It is possible to generalise that technological and transport determinants influence each other and also correspond positively with the overall determinants of interaction. Relationship over a period of five years registered significant variations. Degree of deviation reduced considerably in 1984 as compared to 1979. With the result, technological and transport determinants registered nominal decline in their correlation values (Table 6.8). On the contrary, there has been a nominal increase in the correlation values of the institutional determinants over the two points of time. The value of the coefficient of correlation between the groups of technological determinants and the groups of overall determinants was +0.80, between transport and communication determinants and overall determinants +0.86 and between the institutional determinants and overall determinants +0.34. There has been variations in the inter-correlation as well.

TABLE 6.8: Correlation Matrix of Determinants in Meerut — 1979

Variable	*(A) Mean*	*(B) Standard deviation*
VAR001	0.9989	0.4795
VAR002	0.9983	0.5487
VAR003	1.0006	0.2820
VAR004	2.9978	1.0515

(C) Coefficient of Correlation

	VAR001	*VAR002*	*VAR003*	*VAR004*
VAR001	1.0000 (0) S=0.001	0.8783 (18) S=0.001	0.0880 (18) S=0.364	0.9378 (18) S=0.001
VAR002		1.0000 (0) S=0.001	0.0277 (18) S=0.457	0.9299 (18) S=0.001
VAR003			1.0000 (0) S=0.001	0.3237 (18) S=0.095
VAR004				1.0000 (0) S=0.001

VAR001 = Technological determinants
VAR002 = Transport determinants
VAR003 – Institutional determinants
VAR004 = Overall determinants
(Coefficient/(Cases)/Significance)

Correlation between technological and transport and communication determinants was 0.47 in 1984 compared to 0.87 in 1979. Similarly, correlation between technological and institutional determinants was +0.27 in 1984 (Table 6.9) compared to +0.09 in 1979. On the contrary, relationship between transport and communication determinants and institutional determinants was –0.04 in 1984 compared to +0.03 in 1979. These facts explain that institutional determinants are gradually gaining significance over other determinants. It is, perhaps, due to fast declining relationship between the overall determinants and transport and communication and technological determinants and the gradual increase in the relationship of overall and institutional determinants. As such prevailing modes of transport and communication along with the production-technology require significant modifications to bring about desired changes in the present level of interactions. It is true that growing pressure of interaction cannot be efficiently carried out by the existing modes of transport and technology. Institutional determinants are gaining significance in accelerating the process of interaction. These determinants are expected to exert greater influence on the interaction. The observed trend is not very encouraging because institutional growth is of little relevance to the development unless it is accompanied by the same rate of growth in the production process, i.e., agricultural and industrial productions. As such, increasing movement of commodities and people would require changes in the production technology and modes of transport and communications. Some account of these changes have been discussed in the preceding chapters.

Conclusion

The foregoing observations may be concluded as follows:

(i) While Meerut and Daurala have very high scores of technological and transport determinants, Mawana and Chhaprauli have very high scores of technological and institutional determinants. In terms of technological changes as many as eight blocks did not register any change in their earlier levels. However, rest of the 10 blocks registered changes from their earlier levels,

(ii) It is observed that areas registering high scores have a tendency to grow more or less at the rate over a period of time whereas areas with low scores have a tendency to register unequal growth of units. While Kharkhauda-Daurala axis registered high to very high disparity, Jani-Sardhana with the inclusion of Pilana and Binoli blocks observed very low disparity. Blocks of Khekra and Chhaprauli on the west, Parikshhatgarh on the east and Saroorpur and Rohta on the centre registered medium disparity. As against this, Mawana-Hastinapur axis registered low level of disparity,

TABLE 6.9: Correlation Matrix of Determinants in Meerut — 1984

Variable	*(A) Mean*	*(B) Standard deviation*
VAR001	1.0006	0.3043
VAR002	0.9989	0.4936
VAR003	0.9989	0.1870
VAR004	2.9983	0.7318

(C) Coefficients of Correlation

	VAR001	*VAR002*	*VAR003*	*VAR004*
VAR001	1.0000 (0) S=0.001	0.4696 (18) S=0.025	0.2665 (18) S=0.143	0.8006 (18) S=0.001
VAR002		1.0000 (0) S=0.001	–0.0385 (18) S=0.440	0.8600 (18) S=0.001
VAR003			1.000 (0) S=0.001	0.3403 (18) S=0.084
VAR004				1.0000 (0) S=0.001

VAR001 = Technological determinants
VAR002 = Transport determinants
VAR003 = Institutional determinants
VAR004 = Overall determinants
(Coefficient/(Cases)/Significance)

(iii) While correlation with overall determinants was maximum in case of technological determinants ($r = 0.94$); it was minimum in case of institutional determinants ($r = 0.32$). This relationship was 0.93 in case of transport and communication determinants. Over a period of time, degree of relationship with overall determinants registered marginal decline in case of technological ($r = 0.80$) as well as transport and communication determinants ($r = 0.84$), whereas it registered increase in case of institutional determinants ($r = 0.34$). It is, thus, established that factors of transport and communications are the most important explanatory determinants of interactions, and

(iv) As a group of determinants, transport and communication factors explain interaction (flow of commodities and human movement) to the extent of 43 per cent. Transport and communication factors along with the technology as joint factors (step II) explain interactions up to 60 per cent. Institutional determinants in association with the factors of transport and communications and technology explain to the extent of 69 per cent of the total interaction. Time as a factor of change explains interaction up to five per cent. Thus, about three-fourth of the interaction is explained by the three groups of determinants over a period of time. About one-fourth of the interaction still remains to be explained. It is, therefore, obvious that other factors, besides those considered in this exercise, are also significant in explaining the extent and level of interaction in district Meerut.

7

PROBLEMS AND PROSPECT OF INTERACTION

INTRODUCTION

THIS CHAPTER is aimed at analysing the gravity of the problems and the extent of the prospects of interactions in district Meerut. Since problems denote constraints and prospects denote possibilities, evaluation of these issues becomes necessary to devise ways and means for the interactions. Plans at different hierarchical levels are aimed at raising the growth rates by resolving and reducing the constraints. Lack of correspondence between issue and plans leads to the accentuation of inequalities and the problems. As a consequence, decline in the development is observed which deters prospects of interaction greatly. Apparently this chapter is sub-divided into two parts for the convenience of the analysis. The first part relates to the study of problems. Here set of problems are supposed to be constituting three main dimensions, viz., geographical, social and economic. Besides these dimensions, overall problems have also been calculated. A few problems, which could not be quantified due to non-availability of data, have been discussed in brief under other problems. The second part is devoted to the measurement of the sectoral development, plan achievements and the evaluation of the prospects of interaction. On the assumption that interaction emanates from the processes of production and planning, an attempt has been made to analyse the prospects of interaction. The process of production and planning subsequently leads to the rise in the flow of commodities and human movements.

Exercise is based on the assumption that all indicators are equally important and that problems and prospects are not the same every where (i.e., they vary over the space). It has been worked out by using 34 indicators explaining the gravity of the problems and 28 indicators explaining extent of the prospects of interaction. Values of the indicators have been normalised for the purposes of addition and comparison. Values of the each set of indicators were divided by their respective means. This method helps in reducing the values considerably and keeps the data distribution unchanged. If data distribution is normal, values are supposed to be near one. Similarly, groupwise values of the indicators have jointly formed the basis for mapping the overall problems and prospects of interaction. To explain the spatial patterns of problems and prospects, maps have been drawn by applying choropleth technique. Each part of the chapter is discussed below using list of indicators, methods used and results derived:

PART I : PROBLEMS

This part of the Chapter presents the nature of problems, their identification and class determination over different blocks of the districts. An attempt has been made to measure the maximum possible problems under three groups, viz., geographical, social and economic. As all problems could not be measured directly, a few of them have been measured through proxy parameters. Most of the information used in these indicators relates to the period 1984. However, information about population refers to the periods of 1971 and 1981.

Indicators Used

The assessment of the problems has been made with the help of 34 indicators. Of these 14 indicators belong to the geographical problems, eight to the social problems and 12 to the economic problems. Following is the list of the indicators used to measure the problems:

Geographical Problems

(i) Distance of block headquarters from Meerut city (the district headquarters),

(ii) Distance of nearest railway station from block headquarters.
(iii) Percentage of area under forest, *banjar* and fallow lands to total area,
(iv) Proportion of villages over five km distance from metalled road,
(v) Proportion of villages over five km distance from railway station/halt,
(vi) Proportion of villages over five km distance from bus stand/stops,
(vii) Proportion of villages over five km distance from the post office,
(viii) Proportion of villages over five km distance from the telegraph office,
(ix) Number of villages over five km distance from the primary health centre,
(x) Number of villages over five km distance from the family planning centre,
(xi) Number of villages over five km distance from the animal hospital,
(xii) Number of villages over five km distance from the artificial semen centre,
(xiii) Number of villages over five km distance from the agricultural credit societies, and
(xiv) Number of villages over five km distance from the block.

Social Problems

(i) Percentage growth of population during 1971-81,
(ii) Density of population per sq. km,
(iii) Share of illiterate population to the total population,
(iv) Share of over five hectares of land-holding to total agricultural land,
(v) Dispersal (i.e., minus deviation only) index of piggery units from the district average,
(vi) Dispersal index of milk societies from the district average,
(vii) Proportionate share of scheduled caste students to the total scheduled caste population, and
(viii) Dispersal index of agricultural service centre from the district average.

Economic Problems

(i) Proportion of unirrigated area to total area,
(ii) Dispersal index of industrial worker from the total worker,
(iii) Number of villages over five km distance from marketing societies,
(iv) Number of villages over five km distance from animal bazars,
(v) Number of villages over five km distance from seed godowns,
(vi) Number of villages over five km distance from fertilizer depots,
(vii) Number of villages over five km distance from bazars,
(viii) Number of villages over five km distance from wholesale *mandis,*
(ix) Number of villages over five km distance from fair price shops,
(x) Number of villages over five km distance from land development bank,
(xi) Dispersal index of full-time workers from the total workers, and
(xii) Concentration index of marginal workers from the total workers.

Methods Used

Methods used in this part of the chapter relate to the identification of problems, their classification and the measurement of the intensity of the problems. Problems of the district and its constituent units have been worked out in the following successive steps:

Step I : *Identification*

Identification of the problem is a difficult exercise as there are numerous problems in the district and they are not the same every where. As such, it was decided to identify the problems concerned with the interaction. Data about each problem were neither available through office records nor was it feasible to collect through field survey. As such some proxy indicators have been used to get an idea of the problems and their gravity in case of district Meerut.

Step II : *Classification*

The problems have been classified on the basis of their nature into three groups, viz., geographical, social and economic. Natural as well as environmental constraints, such as, floods, wastelands and physical distances, have been grouped under geographical problems. Similarly, growing pressure of population, illiteracy and concentration of backward population have been grouped under social problems. Growing dependency, lack of economic diversity, inadequate infrastructure and lack of institutional feed-back have been grouped under economic problems.

Step III : *Measurement of the Problems*

To measure the relative strength of various units, values of the indicators have been arranged in an ascending order and are marked with the position of average. Three categories are formed as follows:

(i) More than the average explains concentration,
(ii) Average or near average explains normal distribution, and
(iii) Less than the average explains dispersal or a position below average.

Problems are, therefore, inversely proportional to the development. For instance, reach to a facility becomes a problem, if it involves a distance more than the normal. Distribution below normal indicates deficiency or lack of the normal stimulii. As such distribution below normal, i.e., minus deviation refers to the dispersal index. Greater the value of minus deviation, higher is the index of dispersal. Levels of problems have been identified on the basis of perceptible gaps in the distribution as intense, moderate, low and very low problems. While intense problems correspond with maximum minus deviation, very low problems correspond with maximum positive deviations. As such index of development is an inverse index for problems.

Following formula has been used to calculate the index of problems.

$$IP = 100 + (\frac{1 - \frac{x}{\bar{x}}}{100}$$

wnere,

IP = Index of problems.
X = Values of the indicator X.
$\bar{X}$ = Mean value of the indicator X.

Results Derived

Results derived through the above procedure have been explained groupwise. Separate account for each group of the problems is given below:

Geographical Problems

Geographical problems (Table 7.1) are associated with the issues of regional environment. Attempts to transform environment as per human needs have created imbalances in the natural environment expressing presence of geographical problems. Geographical problems, therefore, pose the basic questions to the regional development. District Meerut is one of the most developed districts of the country and is characterised by levelled land, fertile soil and assured water supply for irrigation. However, there are some geographical problems as well which need to be solved to achieve still higher rates of agricultural growth. The most pressing geographical problems in the district are floods, wastelands and receding underground water level. Growing menace of these problems poses serious challenge to the development. Besides these problems, distance is also a crucial factor influencing intensity and extent of interaction. Since development is inversely proportional to the distance, from growth nucleus, areas away from it experience little impact of growth impulses. Modern means of communications and modes of transport have reduced the distance considerably. However, there are a few areas which have low to very low level of connectivity. Due to lack of efficient linkages, level of interaction (economic as also social) remains very low. The index value of geographical problems ranged between 0.63 (Sardhana) to 1.79 (Parikshhatgarh). Intense geographical problems were recorded in four blocks. Three out of four blocks are situated along the Ganga and the Hindon rivers. In addition to Saroorpur and Kharkhauda blocks, south-western part of the district recorded moderate problems. Blocks of Pilana, Jani, Rohta, Daurala and Rajpura formed a cluster of low geographical problems whereas Baraut-Chhaprauli on the north-west and Meerut-Machhra in the centre registered very low level of geographical problems. It is, therefore, clear that the intensity of geographical problems is proportional to the distance from the Meerut city – the district headquarters – and the regional node for all types of interactions. Further, geographical problems are intense in the areas

TABLE 7.1: District Meerut — Intensity of Problems — 1984

Blocks	Problems											
	Geographical			Social			Economic			Overall		
	Value	Rank	Class	Value	Rank	Class	Value	Rank	Class	Value	Rank	Class
1. Meerut	0.80	14	VL	1.05	4	I	0.96	9.5	L	2.81	10.5	L
2. Jani	0.86	11.5	L	0.93	16	VL	0.87	13.5	VL	2.66	14	VL
3. Rohta	0.84	13	L	0.96	10	L	0.84	15	VL	2.64	15.5	VL
4. Rajpura	0.90	9.5	L	0.90	18	VL	0.81	17	VL	2.51	17	VL
5. Kharkhauda	1.03	6	M	1.01	7	M	0.93	12	L	2.97	9	M
6. Mawana	1.11	4	I	0.94	14	VL	0.94	11	L	2.99	7	M
7. Hastinapur	1.68	2	I	0.95	11.5	L	1.42	1	I	4.05	1	I
8. Parikshhat-Garh	1.79	1	I	0.98	9	M	1.14	4	I	3.91	2	I
9. Machhra	0.69	17	VL	0.94	14	VL	0.77	18	VL	2.40	18	VL
10. Sardhana	0.63	18	VL	1.18	1	I	0.87	13.5	VL	2.68	12.5	L
11. Saroorpur	0.99	7	M	0.92	17	VL	1.07	5.5	M	2.98	8	M
12. Binoli	1.34	3	I	1.03	6	M	1.27	2	I	3.64	3	I
13. Daurala	0.86	11.5	L	1.00	8	M	0.82	16	VL	2.68	12.5	L
14. Baghpat	1.10	5	M	0.94	14	VL	1.20	3	I	3.24	4	I
15. Baraut	0.78	15	VL	1.04	5	M	0.99	8	M	2.81	10.5	L
16. Chhaprauli	0.73	16	VL	0.95	11.5	L	0.96	9.5	L	2.64	15.5	VL
17. Khekra	0.96	8	M	1.17	2	I	1.07	5.5	M	3.20	5	M
18. Pilana	0.90	9.5	L	1.13	3	I	1.06	7	M	3.09	6	M

Note. Further details about the variables *see* Appendices 4A, 4B and 4C.

along rivers. As against this, geographical problems are extremely low near Meerut. Areas with urban industrial development usually have low geographical problems and high intensity of interaction. Blocks of Baraut and Chhaprauli present exception to this generalisation. These blocks located in the Yamuna-Hindon tract are distant from Meerut.

Social Problems

Social problems emanate from increasing pressure of population, extent of high illiteracy, lopsided land distribution and localisation of industrial units. Social problems pose relatively slow but persistent threat to the development and as such to the intensity of interaction. Intensity of social problems in the district Meerut ranged between 0.90 (Rajpura block) to 1.18 (Sardhana block) scores. Intense social problems were recorded in the blocks of Khekra, Pilana, Meerut and Daurala; moderate social problems in the blocks of Báraut, Binoli, Daurala, Kharkhauda and Parikshhatgarh; low social problems in the dispersed pockets of Chhaprauli, Rohta and Hastinapur and very low social problems in the blocks of Baghpat, Saroorpur, Jani, Mawana, Rajpura and Machhra. Intense social problems have cropped up due to the demographic and economic imbalances. These problems were recorded as the adverse side effect of urban industrial process in the blocks of Meerut and Sardhana and poor linkages in the blocks of Khekra and Pilana. Social problems, thus, form a bi-modal situation with 'V' shaped distribution curve in district Meerut. Here areas with moderate development correspond with low to very low level of social problems. On the contrary, areas with high or low development correspond closely with intense social problems. These problems get further intensified due to the exploitation of the society in a developed or backward set-up. Moderate development is the critical stage in which intensity of social problems remains minimum. Moderate development refers to normal pressure of population, the average level of education and the balanced sectoral growth of the economy. This situation has high scores of interaction without causing corresponding effects over social problems.

Economic Problems

These problems crop up due to the concentration of landless and marginal farmers, low level of irrigation, mechanisation, industrial participation and distant location of (markets, banks and other) facilities and infrastructure. Economic problems, thus, act as constraints in the process of development. These problems ranged between 0.77 (in block Machhra) to 1.42 (in block Hastinapur) in the district. Relatively large range of economic problems have passed the most serious challenge to the process of development. Table 7.1 shows that variations are relatively more marked in case of economic problems as compared to the social problems. Results reveal that there were intense economic problems in the eastern and western parts of the district, i.e., in the blocks of Parikshhatgarh and Ḥastinapur towards the east and Baghpat and Binoli towards the west. Moderate economic problems were recorded in Baghpat-Binoli complex. While low economic problems were recorded in the blocks of Meerut, Kharkhauda, Chhaprauli and Mawana, very low economic problems were recorded in the blocks near Meerut. Thus, central part of the district had relatively low to very low scale of economic problems compared to the eastern and western parts of the district. It appears that these problems are inversely proportional to the distance from the district headquarters. Due to lack of adequate linkages problems accentuate further. It is, therefore, necessary to solve the economic problems by providing adequate infrastructure and linkages in the various units of the district.

Overall Problems

It is obvious from the foregoing discussion that areas differ in the nature of problems and suffer from their varying influences. Problems combine in an order sequence that varies from one type with the other. It is, therefore, desirable to assess the intensity of all the problems together and to analyse the combination sequence of the problems separately. This set of problems takes a joint stock of the three problems discussed above. The overall problems ranged between 2.40 (Machhra block) to 4.05 (Hastinapur block). Large variations were recorded in the scores of problems by various unit areas. Intense problems were recorded in the blocks of Hastinapur, Parikshhatgarh, Binoli and Baghpat. Moderate problems were recorded in the blocks of Kharkhauda, Mawana, Khekra, Pilana and Saroorpur. While blocks of Meerut, Sardhana and Baraut recorded low level of problems, Rajpura, Machhra, Rohta, Jani and Chhaprauli recorded very low level of problems.

From the foregoing discussion it is evident that most of the problems are either geographical or economic in nature. The two set of problems have a tendency to increase with the distance from Meerut. In contrast to the above, the social problems have an inverse relationship with the geographical problems and with the distance. Social problems, as explained above, present a transitional phase between geographical and economic problems. Social problems are likely to accentuate further with the localisation of development processes in selected areas. Social problems can be solved with the systematic attempts towards decentralised economic planning and the development of manpower. Direction and sequence of the problems vary from one part of the district to the other. While geonomic and social problems prevail in the western part of district, economic, social and geographical problems dominate in the centre. The eastern part of the district is characterised by the sequence of geographical, social and economic problems (Fig. 7.1).

Responses to the Problems of Interaction

To analyse the inter-relationships of the problems, an exercise on correlation coefficient has been worked out. The results of the exercise reveal that geographical problems have a negative correlation (–0.21) with social problems whereas positive correlation with economic (0.76) and overall problems (0.93). Social problems, on the other hand, have insignificant positive relationship with economic (0.05) and overall problems (0.05). It has a negative correlation with geographical problems. Economic problems, on their turn, have extremely high positive correlation with the overall problems (0.91). Results further reveal that geonomic problems exercise the maximum and social problems the minimum influence upon the areal units. Thus, areas faced with the geographical problems also experience economic problems. Possibilities of interaction weaken with the growing menace of geonomic problems. Eastern as well as western margins of the district have relatively low intensity of interaction whereas central part of the district recorded high intensity of interaction despite growing social problems (Table 7.2).

PART II : PROSPECTS OF INTERACTION

Temporal trends of interaction indicate future possibilities. Trends of interaction have been discussed, in detail, in the first four chapters. Apart from recent trends of interaction, production and planning parameters have also been used to analyse the prospects of interaction. While production parameters deal with agricultural as well as non-agricultural interaction, planning parameters deal with the extent of achievements and budget allocations for financial year 1987-88. The overall prospects of interaction depend upon the joint influences of these parameters.

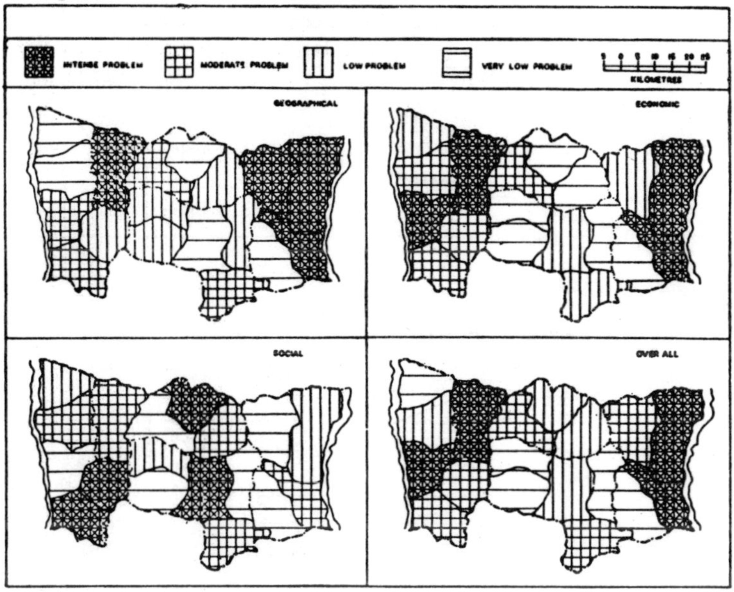

FIG. 7.1: Problems of interaction in Meerut.

Indicators Used

Prospects of interaction have been worked out with the help of 28 variables. This exercise includes non-agricultural (six variables), agricultural (six variables), planning (nine variables) and the allocational (seven variables) factors supposed to be influencing the prospects of interaction.

Following is the list of variables for prospects of interaction:

Non-agricultural Development

(i) Degree of connectivity,

(ii) Frequency of train halts and bus stops,
(iii) Number of *Gur* and Jaggarey *kolhus* per 10 villages,
(iv) Number of cane-crushers per 10 villages,
(v) Percentage of households benefited (credit) for industries under integrated rural development programmes, and
(vi) Percentage of households benefited for service and trade under integrated rural development programmes.

TABLE 7.2: Variable Response to the Problems of Interaction — 1984

Variable	*(A) Mean*	*(B) Standard deviation*
VAR001	0.9994	0.3168
VAR002	1.0011	0.0849
VAR003	0.9994	0.1740
VAR004	3.0000	0.4609

(C) Coefficients of Correlation

	VAR001	*VAR002*	*VAR003*	*VAR004*
VAR001	1.0000 (0) S=0.001	-0.2143 (18) S=0.197	0.7663 (18) S=0.001	0.9372 (18) S=0.001
VAR002		1.0000 (0) S=0.001	0.0551 (18) S=0.414	0.0575 (18) S=0.410
VAR003			1.0000 (0) S=0.001	0.9144 (18) S=0.001
VAR004				1.0000 (0) S=0.001

VAR001 = Geographical problems
VAR002 = Social problems
VAR003 = Economic problems
VAR004 = Overall problems
(Coefficient/(Cases)/Significance)

Agricultural Development

(i) Proportion of commercial crops to the net sown area in the block,
(ii) Proportion of vegetable and fodder crops to the net sown area in the block,
(iii) Proportion of area under orchards in a block to total arca under orchards in the district,
(iv) Number of animal wealth per thousand of households,
(v) Number of piggery units per lakh of households, and
(vi) Credit-deposit ratio (amount of credit per unit of deposits).

Integrated Rural Development Programmes and Poverty Eradication

(i) Proportion of workers to the total population,
(ii) Number of repair units per thousand of households,
(iii) Number of hospital beds per ten thousand of population,
(iv) Proportion of households benefited under IRDP to total houscholds in thc block,
(v) Proportionate share of the households above poverty line to the total households in the block,
(vi) Rate of poverty eradication in the block,
(vii) Pattern of IRDP beneficiaries among rural workers,
(viii) Pattern of financial support for sectors of rural economy, and
(ix) Nature of IRDP financial support to the rural poor households.

Budget Allocation in Annual Plan

(i) Proportion of budget allocation (1987) for agriculture and allied activities,
(ii) Proportion of budget allocation (1987) for small scale industries,
(iii) Proportion of budget allocation (1987) for tertiary activities,
(iv) Proportion of budget allocation (1987) for housing sector,
(v) Proportion of budget allocation (1987) for education,
(vi) Proportion of budget allocation for consumption loan to the weaker section of the society, and
(vii) Proportion of overall budget (1987) of a block to the total district budget.

Methods Used

Based on the above indicators an attempt has been made to measure the prospects of interaction. Prospects of interactions are assumed to be proportional to the development and planning. While development is explained through agricultural and non-agricultural groups of indicators, planning is explained by groups of indicators related to the IRDP, poverty eradication and the budget allocation. Normalised values of the indicators have been calculated by dividing values with their respective means. Values, thus obtained, were added to get the relative position of each group of indicators. Having obtained the relative position of the indicators it became necessary to work out a composite index for measuring the prospects of interaction. Prospects of interaction have been derived through the cumulative strength of these groups of indicators. Range of the indicators is divided into five orders of very high, high, medium, low and very low. These orders have helped in the identification of the level of a unit and its spatial linkages. By applying this procedure, prospects of interaction have been calculated and explained.

Method is given as under :

$$PI = X \div \frac{\Sigma X}{N}$$

where,

PI = Prospects of interaction.
X = Value of the indicator in unit *X*.
N = Number of observations.

Results Derived

Detailed account of the broad groups of indicators are given below:

Levels of Non-agricultural Development

Factors, such as, transport, trade, commerce, facilities and services, besides industries, form the basis for non-agricultural development in the region. In the growing process of urban-industrial development, growth of non-agricultural sector becomes most important. The index value of (Table 7.3) non-agricultural development ranged between 0.71

(Daurala block) to 1.49 (Saroorpur block). Indices of very high development were recorded in the blocks of Chhaprauli, Saroorpur, Sardhana and Kharkhauda. The index values ranged between 1.11 to 1.49. High scores of non-agricultural development were recorded in the blocks of Rohta, Machhra and Binoli. The index value ranged between 1.03 to 1.07. Figure 7.2 shows a direct correspondence between the areas registering high agricultural development and the areas registering high non-agricultural development. Areas registering moderate scores of non-agricultural development have clusters near areas of high development. The index values ranged between 0.96 (Hastinapur) to 1.11 (Kharkhauda). Blocks of Parikshhatgarh, Baraut and Meerut recorded low level of non-agricultural development with a score range between 0.86 to 0.96. There were very little variations in the scores of the blocks within this category. Further, scores of very low level of non-agricultural development ranged between 0.71 to 0.76. Blocks with low level of development formed a cluster in the south-western part of the district covering blocks of Pilana, Khekra and Baghpat and with an isolated block in the north-central part of the district. It is evident that the level of non-agricultural development corresponds closely with the agricultural development. While Binoli-Sardhana complex registered high to very high index of development both in agricultural as well as non-agricultural sectors of economy; Pilana-Khekra axis registered very low level of development in both the sectors. However, inverse relationship existed in cases of Chhaprauli and Baghpat blocks. The former recorded very high scores in non-agricultural sector whereas the latter recorded higher scores in agricultural sector.

The results reveal a positive correspondence between the two sectors of economy (r being 0.90). Since agriculture provides the base for non-agricultural development; sectors of economy become inter-dependent and positively associated. Some of the agricultural products are used as raw material base for local processing units in the district. The most common example is that of crane-crushers. *Kolhus*, expellers and flour mills. This is a unique functional dependency in which one sector offers the feed-back to the other. Contrary to this, non-agricultural development has limited links with the agricultural development in other parts of the country.

TABLE 7.3: District Meerut — Sectoral Development, Planning and Prospects of Interaction — 1984

Blocks	Prospects														
	Non-agricultural			Agricultural			I.R.D.P. and poverty Eradication			Budget allocation in Annual Plan (1987)			Overall prospects of Interaction		
	Value	Rank	Class	Value	Rank	Class	Value	Rank	Class	Value	Rank	Class	Value	Rank	Class
1. Meerut	0.90	12	L	0.92	11.5	M	2.65	1	VH	5.08	1	VH	9.55	1	VH
2. Jani	0.97	10	M	0.64	17	VL	1.23	5	H	0.70	12.5	L	3.54	11	M
3. Rohta	1.07	5	H	0.93	10	M	1.26	4	VH	0.57	18	VL	3.83	7	H
4. Rajpura	1.00	8	M	0.71	16	VL	0.78	11	M	0.69	14	L	3.18	16	VL
5. Kharkhauda	1.11	4	VH	0.96	7	H	0.51	15	VL	0.67	17	VL	3.25	15	VL
6. Mawana	0.99	9	M	0.85	13	L	0.62	14	L	0.89	4	VH	3.35	13	L
7. Hastinapur	0.96	11	M	0.92	11.5	M	1.30	2.5	VH	0.82	6	H	4.00	4	VH
8. Parikshhat-Garh	0.89	13	L	0.99	5	H	0.57	16	VL	0.72	10.5	M	3.17	17	V!
9. Machhra	1.03	7	H	1.19	2	VH	0.89	10	M	0.74	8	M	3.85	6	H
10. Sardhana	1.23	3	VH	0.97	6	H	0.47	17	VL	0.94	3	VH	3.61	9	M
11. Saroorpur	1.49	1	VH	0.95	8.5	M	0.64	12.5	L	0.72	10.5	M	3.80	8	M
12. Binoli	1.06	6	H	2.69	1	VH	1.13	9	M	0.79	7	H	5.67	2	VH
13. Daurala	0.71	18	VL	0.95	8.5	M	1.18	7	H	0.73	9	M	3.57	10	M
14. Baghpat	0.72	17	VL	1.06	3	VH	1.20	6	H	0.88	5	H	3.86	5	H
15. Baraut	0.86	14	L	1.02	4	VH	0.45	18	VL	1.00	2	VH	3.33	14	L
16. Chhaprauli	1.46	2	VH	0.84	14	L	1.30	2.5	VH	0.67	15.5	VL	4.27	3	VH
17. Khekra	0.76	15.5	VL	0.58	18	VL	0.64	12.5	L	0.67	15.5	VL	2.65	18	VL
18. Pilana	0.76	15.5	VL	0.83	15	VL	1.17	8	M	0.70	12.5	L	3.46	12	L

Note: Further details about the variables *see* Appendices 5A, 5B, 5C and 5D.

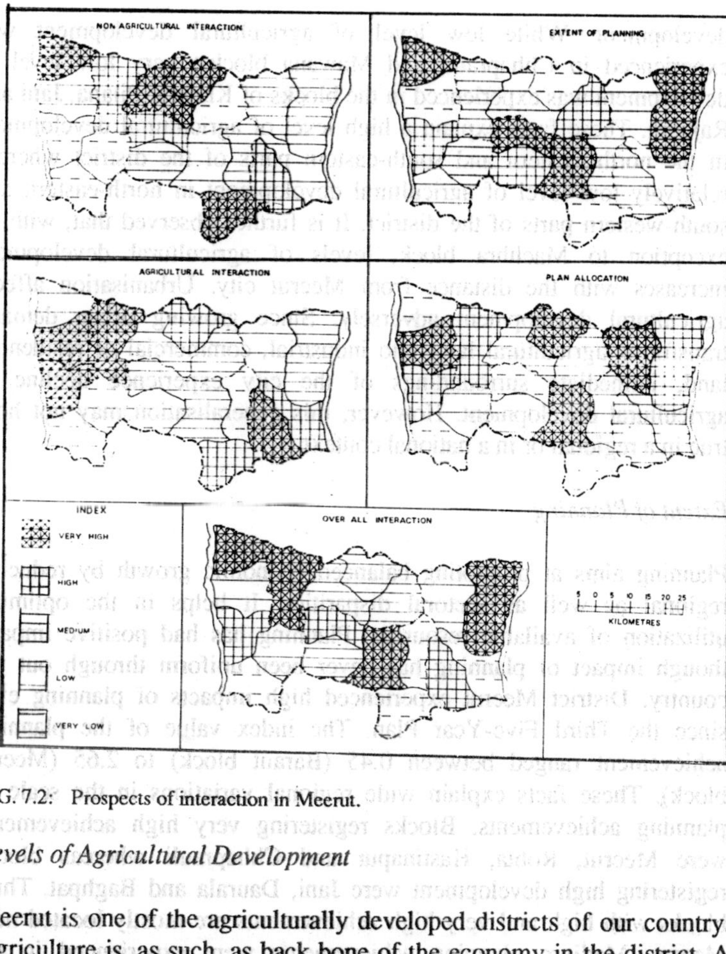

FIG. 7.2: Prospects of interaction in Meerut.

Levels of Agricultural Development

Meerut is one of the agriculturally developed districts of our country. Agriculture is, as such, as back bone of the economy in the district. A separate account of agricultural development is needed to analyse the prospects of agricultural interaction. Results reveal that agricultural development ranged between the score points 0.58 (Khekra block) and 2.69 (Binoli block). Marked variations were recorded at all levels of agricultural development. Very high agricultural development was recorded in the blocks of Baghpat, Baraut and Binoli in the west and Machhra block in the east. Blocks of Kharkhauda, Parikshhatgarh and Sardhana recorded high agricultural development whereas blocks of Saroorpur, Rohta, Meerut and Hastinapur recorded medium development. While low level of agricultural development was experienced in Chhaprauli and Mawana blocks; very low level of development was experienced in the blocks of Khekra, Pilana, Jani and Rajpura. These facts explain a high level of agricultural development in the north-western and south-eastern parts of the district whereas relatively low level of agricultural development in north-eastern and south-western parts of the district. It is further observed that, with an exception to Machhra block, levels of agricultural development increases with the distance from Meerut city. Urbanisation affects agricultural development adversely. Since growing urban demand transforms agricultural land into industrial, commercial or residential land, immediate surroundings of the city experience decline in agricultural development. However, this generalisation may not hold true in a regional or in a national context.

Extent of Planning

Planning aims at promoting balanced economic growth by reducing regional as well as sectoral disparities. It helps in the optimum utilization of available resources. Planning has had positive impact, though impact of planning has never been uniform through out the country. District Meerut experienced high impacts of planning ever since the Third Five-Year Plan. The index value of the planning achievement ranged between 0.45 (Baraut block) to 2.65 (Meerut block). These facts explain wide regional variations in the scale of planning achievements. Blocks registering very high achievements were Meerut, Rohta, Hastinapur and Chhaprauli whereas blocks registering high development were Jani, Daurala and Baghpat. Thus, blocks with high and very high achievements are mostly located near Meerut. Medium planning achievements were experienced in the blocks of Binoli and Pilana along the river Hindon and Rajpura and Machhra near Meerut city. While blocks of Khekra, Saroorpur and Mawana experienced low planning achievements; blocks of Baraut, Sardhana, Parikshhatgarh and Kharkhauda experienced very low level of achievements. These facts lead to the following generalisations.

High to very high planning achievements are experienced in the blocks near Meerut. Extent of planning achievement corresponds closely with the level of agricultural as well as non-agricultural development. Planning achievements have been recorded higher both in the areas of high as well low levels of development. While developed areas experienced higher achievements to maintain their rising growth trend, backward areas experienced high planning achievements to achieve the level of regional average. Planning achievement registered a bi-model trend converging at the average. Planning is, thus, faced with a situation of a dichotomy where attention is being paid towards areas of high and low levels of development. While areas with high development respond quickly, areas with low development respond slowly to the packages of planning. Allocation of budget is greatly influenced by the areas of scarcity or low development to accelerate the pace of progress in backward areas. Similarly, urban-industrial areas demand a higher proportion of budget allocation for their requirements of housing, industrial complexes, services and institutions. The current trend of planning lacks sincere implementation. In addition to that, allocation of the budget does not correspond with the priorities and problems of the region. In the absence of desired co-ordination and sincere implementation, disparities grow unchecked causing lopsided sectoral growth in different areal units.

Plan Allocation

Plan allocation is regarded as one of the dimensions influencing prospects of interaction in the region. Annual plan 1987-88 for Meerut has been approved by the agencies of development and planning in the district. A sum of Rs. 272.89 million is proposed to be spent on various sectors during 1987-88. Budget allocation explains areal as well as sectoral gaps in the district. Sectoral allocation of the budget reveals top priority for agriculture and allied activities. A sum of Rs. 100 million is allocated for this sector. Small scale industries rank second in the budget allocation and accounted for a sum of Rs. 93 million. Tertiary sector activities ranked third in the budget allocation, the amount of budget being Rs. 47 million. Besides these major sectoral allocations, minor allocations have also been proposed. Minor allocations included the sectors, like, education, housing and consumption loan for weaker sections. A sum of Rs. 2.1 million is allocated for housing, Rs. 1.7 million for education and Rs. 1.6 million for consumption loan. Spatial allocation reveals nearly 31 per cent of the district budget for Meerut block. Remaining 69 per cent of the budget is shared by rest of the 17 blocks. While Baraut block accounted for 5.55 per cent of the budget; blocks of Baghpat,

Sardhana, Mawana, Binoli and Daurala accounted for four to five per cent each. About three to four per cent of the budget for each block is allocated for remaining 11 blocks of the district. It is, thus, evident that there are marked regional as well as sectoral variations in the allocation of the budget for 1987-88.

The present budget is, therefore, likely to increase regional as well as sectoral disparities. The index values of the budget allocation ranged between 0.57 (Rohta block) to 5.08 (Meerut block). Areas registering very high budget allocation are Meerut, Mawana, Sardhana and Baraut; high budget allocation for Hastinapur, Binoli and Baghpat; medium plan allocation for Daurala, Machhra and Parikshhatgarh; low plan allocation for Jani, Rajpura and Pilana and very low plan allocation for the blocks of Rohta, Kharkhauda, Chhaprauli and Khekra. Plan allocation is influenced by the extent of urbanisation. Amount of the budget decreases with increasing rural ratio in an area. Commodity flows and human movements are likely to be adversely affected by the growing disparity. Since sizable proportion of the budget is allocated for Meerut, prospects of interaction are generally higher.

Overall Prospects of Interaction

Prospects of interactions have been worked out by assessing the relative importance of each unit in terms of agricultural as well as non-agricultural development and achievement of planning and the budget allocation for the plan. Based on the assumption that prospects of interaction are influenced by the economy, planning and the budget, composite scores of these factors have been worked out to explain the relative importance of individual units. Units with almost similar prospects have been grouped together to explain the prospects of interaction. Prospects of interaction ranged between 2.65 (Khekra block) to 9.55 (Meerut block) in district Meerut. There are marked variations in the prospects of individual units. Results reveal very high prospects of interaction for the blocks of Meerut, Binoli, Chhaprauli and Hastinapur; high prospects of interaction for the blocks of Baghpat, Machhra and Rohta and medium prospects of interaction for the blocks of Jani, Sardhana, Saroorpur and Daurala. While low prospects of interaction are revealed for the blocks of Pilana, Mawana and Baraut; very low prospects of interaction are calculated for the blocks of Kharkhauda, Rajpura, Parikshhatgarh and Khekra. It is established that Meerut-Chhaprauli, Meerut-Baghpat and Meerut-Hastinapur corridors have very high prospects of interaction; blocks near Meerut – medium to high prospects and scattered pockets between these two types of areas – low to very low prospects of interaction. Prospects of interaction are usually higher towards the periphery compared to the central part of the district. These areas are endowed with rich agricultural resources. However, existed resources need not be optimally utilised due to constraints on agricultural development. As such, eastern and western margins of the district have high prospects of interaction if constraints are removed. Meerut city will continue to maintain the regional primacy of the interaction due to low geonomic constraints and growing functional diversity. The range of average daily interaction, which is approximately 75 km now, is expected to extend further up to a range of approximately 100 km by the turn of the next plan period. It must be visualised that pressure of population will accentuate further in urban-industrial areas of the district that would need greater emphasis on transport planning for inter-regional connections. Growing regional inequality poses serious challenge to the prospects of overall interactions.

Responses to the Prospects of Interaction

To work out the response of development on the prospects of interaction, coefficients of correlation have been calculated. Of the four groups of development, budget allocation recorded the maximum response to the prospects of interaction, r being 0.91. It is closely followed by the achievements of planning (I.R.D.P. and poverty eradication programmes) which recorded a correlation value of 0.84. On the contrary relatively low correlation values are recorded in cases of agricultural (r=0.30) and non-agricultural development (r=0.035). Negative correlations have been observed between non-agricultural development and planning achievements and between agricultural development and the budget allocations (Table 7.4). It is, thus, clear that prospects of interaction rest on the prosperity of the individual which depends on the achievement of planning and the budget allocation (I.R.D.P. and poverty eradication). Thus, process of regional development and planning accelerates the prospects of interaction in a region. Prospects of interactions would be low if planning achievements and budget allocations are not directly linked with the resources and problems of an area. Contrary to this, prospects would be higher if planning is linked with the priorities and the resources of an area. Since sizable section of our population lives below poverty line, prospects of interaction would depend upon the I.R.D.P. programmes and their implementation. Unless this section of the society is uplifted to the average level of prosperity, prospects of interaction would not be high. In the absence of people's participation from all walks of life, interaction remains a phenomenon of economically rich section of the society. This section of the society is already in the orbit of interaction and as such enjoys high prospects of interaction. Disparity needs to be removed to observe balanced interaction between and within the regions. Results reveal that I.R.D.P. and poverty eradication programmes have positively influenced the prospects of interaction in district Meerut. Here, inter-regional disparity registered a decline which has consequently raised the prospects of interaction. Planned efforts are required to raise the level of poor to have an overall high prospects of interaction in the district (Table 7.4).

Conclusion

Discussion about problems and prospects of interaction leads to the following generalisations.

(i) Problematic areal units have positive prospects of interaction. This is true in cases of Binoli and Hastinapur where problems are intense due to floods, wastelands, poor linkages and low institutional intensity. In the absence of the adequate bridges over major rivers (the Ganga and the Yamuna) and bypasses around major urban centres, inter-regional flows and human movements get diverted to different areal units. Extremely high prospects of interaction lie ahead, if the problems are solved,

(ii) Developed blocks (Meerut, Chhaprauli, Daurala and Mawana) have high scores of interaction. Future prospects of interaction would be proportional to the changes brought in the present level of production technology and transport planning. While intensity of interaction is inversely proportional to the intensity of problems; prospects of interaction correspond positively to the agricultural as well as urban-industrial development. Process of development and planning should, therefore, be directly linked with the resources and problems of the area,

TABLE 7.4: Correlates of the Prospects of Interaction — 1984

Variable	*(A) Mean*	*(B) Standard deviation*
VAR001	0.9983	0.2234
VAR002	1.0000	0.4465
VAR003	0.9994	0.5210
VAR004	0.9989	1.0244
VAR005	3.9967	1.5192

(C) Coefficients of Correlation

	VAR001	*VAR002*	*VAR003*	*VAR004*	*VAR005*
VAR001	1.0000	0.0908	-0.1567	-0.1251	0.0356
	(18)	(18)	(18)	(18)	(18)
	P=xxxxx	P=0.360	P=0.267	P=0.310	P=0.444
VAR002		1.0000	0.0472	-0.0246	0.3068
		(18)	(18)	(18)	(18)
		P=xxxxx	P=0.426	P=0.461	P=0.108
VAR003			1.0000	0.7627	0.8481
			(18)	(18)	(18)
			P=xxxxx	P=0.000	P=0.000
VAR004				1.0000	0.9102
				(18)	(18)
				P=xxxxx	P=C.000
VAR005					1.0000
					(18)
					P=xxxxx

VAR001 = Non-agricultural development
VAR002 = Agricultural development
VAR003 = I.R.D.P. and poverty eradication
VAR004 = Budget allocation
VAR005 = Overall prospects
(Coefficient/(Cases)/Significance).

(iii) Because of low problems, Meerut has very high prospects of interactions. It is mainly due to increasing industrial and institutional concentrations over here. While geonomic problems have higher relationship with each other, social problems have insignificant relationship with geonomic problems. Consequently, prospects of interaction are inversely proportional to the geonomic problems It is, therefore, essential to remove the socio-economic and environmental bottlenecks to observe high prospects of interaction,

(iv) Intensity of interaction is accelerated through alternate and efficient modes of transport and communications. Production and transport technology requires change (in cases of Meerut, Mawana, Machhra and Daurala blocks) to widen the prospects of interaction,

(v) Institutional growth is essential for increasing communication between urban and rural; agricultural and industrial sectors and for providing basic amenities and services. Blocks of Chhaprauli, Pilana, Saroorpur and Kharkhauda would require institutional feedback to increase their prospects of interactions, and

(vi) Budget allocation explains areal as well as sectoral gaps in the district. Nearly 31 per cent of the district annual budget is earmarked in favour of Meerut block alone. On the contrary as many as 11 blocks were considered for a budget of three to four per cent each. Areal disparity in budget allocation is likely to cause uneven prospects of interaction. Sectoral break up of the budget attached top priority for agriculture and allied activities and as such a sum of Rs. 100 million is proposed for this sector. As against this a sum of Rs. 2.1 million is proposed for housing, Rs. 1.7 million for education, Rs. 1.6 million for consumption loan. It is, therefore, apparent that allocation of the budget is not in favour of bridging the imbalanced regional as well as sectoral growth in the district. As a result prospects of agricultural interaction would increase and interaction from various parts of the district would be centre-oriented instead of being dispersed in nature. Meerut city is likely to command a much greater and distant interaction than what was experienced so far. Prospects of interaction have an extremely high positive correlation (r being 0.91). It is closely followed by the achievement of planning (r being 0.84). The present level of agricultural ($r = 0.30$) as well as non-agricultural development ($r = 0.035$) presents low level response to the interaction. Prospects of interaction are, therefore, proportional to the changes in the production technology (agricultural as well as non-agricultural), alleviation of poverty and reduction of income inequality and regional disparity.

8

CONCLUSION

INTERACTION SIGNIFIES the links between the places in a given region. As such, present study unfolds the links of Meerut city at different levels of regional hierarchy. This chapter is based on the observations made in the preceding chapters. Following conclusions emerge from the study:

(i) Increasing input demand and surplus supplies have earned a distinct position for Meerut in regional as well as in national contexts. Goods of established quality are supplied to the national and international markets. In terms of supply, it generates huge quantity of agricultural as well as industrial surplus. Meerut has international links for the supply of sports goods, handloom cloth and scissors. It has a nation-wide supply of band goods, auto tyres, printed books in addition to sports goods, handloom cloth and cutlery. Supply of sugar, medicines, surgical goods, wheat, potato, *khandsari* and *jwar* remains confined to the neighbouring States whereas that of vegetables, fruits, bread, milk, soap, livestock feed, cycle and rickshaw tyres to the neighbouring places in the region.

Similarly, light goods arrive here from distant places whereas heavy goods, industrial machinery and consumption goods of non-local production from neighbouring places. In terms of demand, city recorded receipts of defence goods, construction goods, spices, dry fruits, textiles, agricultural machinery, fertilizer, leather and rubber from different parts of the country, petroleum, coal, gas, soap, detergents, utensils, H.Y.V. seeds and handloom *khaddis* from the northern States and timber, firewood, rice, pulses, electronics, electrical and plastic goods from neighbouring places,

(ii) Results reveal that demand and supply (r = from 0.73 to 0.99) and industrial output and outflow (r = 0.99) hold significant positive association whereas market prices and *mandi* arrival (r = –0.249), agricultural output and outflow (r = –0.93) and distance and flow (–0.42) have significant negative association. Most of the industrial functions tend to concentrate leading to the marked variations in the areal growth whereas agricultural functions tend to be distributed ubiquitously. Since the share of large production units in the overall income is greater than the share of small units together, income inequalities tend to remain high at all levels of hierarchy,

(iii) Intensity of human movement remains considerably high within a radius of 100 km from Meerut city, it falls down considerably over a distance of additional 400 km and finally weakens to an insignificant movement up to a distance of 1,600 km. About 55 per cent of the total human movement was for the short distance, 28 per cent for medium and 17 per cent long distances. Nearly 88 per cent of the human movement is through highways and about 12 per cent through railways. About 19.56 million passengers moved over to 200 places yielding an annual earning of about Rs. 75 million. There is a high degree of negative correspondence between distance and the amount of human movement (r being –0.80). However, extent of negative response is greater in case of the movement through railways (r = –0.79) compared to the highways (r = –0.62).

(iv) Of all the purposes for which interactions are made, social interactions are the most important followed by the movement for work place and for availing various kinds of services. Interactions for economic, political and administrative purposes are relatively weak. Purpose of movement becomes selective and of advanced nature with the increasing distance from Meerut city. As a result, human movements decrease with the distance for most of the purposes. A high degree of negative association (r = more than –0.65) was observed for the purposes like education, legal advice, entertainment and alcoholic drinks. Interaction for trade (–0.28), social (–0.24) and religious (–0.15) purposes recorded low level of inverse relationship. It is, however, important to note that the amount of human movement increases with the distance in cases of political, administrative (r = +0.80) and for the search of jobs (r = +0.39),

(v) Distance plays a vital role in deciding the amount and frequency of human movements. Both the amount and the frequency of human movements decrease with the distance. While the function of distance decay operates in the frequency of daily movement, it fails to operate in remaining frequency groups. Explanation to such a relationship may be that people within a short range of distance can afford to move to the city daily because of the extremely high degree of functional inter-dependence. On the contrary, long range interaction is less frequent due to declining functional inter-dependence and increasing travel cost with the distance. While frequency of daily movement (–0.41) holds negative relationship with the distance, frequency of rate (r = +0.63), yearly (+0.35), half-yearly (+0.25) and monthly movements (+0.26) hold positive relationships,

(vi) It is established that private modes of transport follow the function of distance decay more closely compared to the public modes of transport. Impact of distance in reducing the amount of movement remains high in case of public modes of transport, despite low correlations between the two. Tractors, matadors, bicycles and cars have high degree of negative relationships (r = more than –0.52) with the distance; private buses, *buggi* and *tonga* (r = from –0.41 to –0.48) have moderate level of negative relationships and trains, buses and other modes have low level of negative relationships (between –0.15 to –0.37),

(vii) Intensity of interaction (commodity flows and human movements) is proportional to the prosperity of a place and its people. In terms of development as many as 10 out of 18 development blocks in the district have registered rise in their score values of 1984 over 1979. While blocks of Rajpura, Saroorpur, Daurala and Baraut registered rise in their levels of development; blocks of Kharkhauda, Machhra and Baghpat registered fall from their earlier development levels. Remaining 11 blocks maintained the same level of development. Balanced sectoral growth was observed in case of nine blocks, i.e., half of the total blocks of the district. Blocks of Rajpura, Machhra and Kharkhauda registered relative advancement in institutional determinant whereas Sardhana, Saroorpur, Khekra, Hastinapur, Daurala and Baghpat have relative advancement in non-institutional determinants,

(viii) Sectoral and regional disparities registered an overall decline by 1.88 score points during 1979-84. For the blocks registering very high development, disparity is reduced by 1.84 whereas for the blocks registering moderate development it is reduced by 0.06 score points. Contrary to the above, disparity registered increase by 0.23 score points for the blocks recording very low level of development, by 0.10 score points for the blocks recording low level of development and by 0.02 for the blocks recording high development,

(ix) Determinants of interaction registered positive correlation with technological (0.94), transport (0.92) and institutional (0.32) factors. Technological determinants also registered high positive correlation with the determinants of transport (*r* being 0.87). On the contrary, institutional factors registered low degree of positive correlation with all the three factors. Institutional determinants are, thus, gaining importance over time. It is mainly due to the corresponding decline in the extent of relationship between the overall determinants of interaction and transport and technological determinants,

(x) Extent of backwardness in a region reflects the gravity of problems, which in its turn reflects low level of interactions. It has been proved that problems continue to remain and grow with the process of under development. Moreover, problems operate both in isolation as well as in association. Exercise on the problems of areal interaction reveals that geographical problems have low level of negative correlation ($r = -0.21$) with social problems but high positive correlation with economic ($r = +0.76$) and overall problems of interaction ($r = +0.93$). Contrary to this, social problems have very low level of association with economic ($r = 0.05$) and overall problems ($r = 0.05$) of interaction. Economic problems, on their turn, have very high positive association with the overall problems ($r = +0.91$). Blocks faced with geographical and economic problems have low level of interactions.

(xi) Since sizable section of the population, in this part of the country, is educated and economically advanced, prospects of interactions are usually high. Prospects of interaction are inversely proportional to the distance from the city ($r = -0.43$). Since non-agricultural interactions are mostly associated with the urban centres, they hold a low level of positive relationship ($r = 0.034$) with the distance. As against this, prospects of interactions have relatively positive correlation with the agricultural development ($r = +0.30$). Similarly, achievements of planning reflect higher prospects of interaction ($r = 0.84$). Nearly 31 per cent of the district budget (Rs. 100 million) is allocated for the development of Meerut block alone. The extent of budget allocation varies from four to over 31 per cent in different blocks. It shows the maximum positive correlation with the prospects of interaction (r being 0.91). Disparities (sectoral as well as spatial) are likely to accentuate further creating sharp variations in the income and as such in the interactions. Meerut-Chhaprauli and Meerut-Baghpat zones have very high prospects of agricultural interactions whereas Binoli-Sardhana zone has high prospects of non-agricultural interactions. In addition to the above, Meerut-Machhra and Meerut-Hastinapur zones have moderate to high prospects of overall interaction. The increasing pressure of interaction on Meerut-Delhi, Meerut-Garhmukteswar and Meerut-Muzaffarnagar routes have already posed problems of traffic congestion. A long term transport planning would be required for growing traffic pressure on various roads. Apart from transport and traffic planning, necessary changes would be needed to modernise the production technology and modes of transport to raise the level of commodity flows and human movements. Alternate modes of transport, alternate routes, bypass and bridges have to be planned to avoid inconvenience and delay caused by *enroute* towns and cities. Industrial technology needs modernisation for higher output and outflow whereas agricultural technology needs transfer of productive means at all sections of the rural society.

APPENDICES

APPENDIX 1

Nature of Surplus and Deficit Commodities (1981-83)

(Weight : In Tonnes)

Commodities	*1981*		*1982*		*1983*	
	Surplus	*Deficit*	*Surplus*	*Deficit*	*Surplus*	*Deficit*
1. Printed books	—	316.00	1024.00	—	592.00	—
2. Medicines	—	392.00	30.00	—	—	171.00
3. Sports goods	958.14	—	1417.42	—	1396.84	—
4. Waste cotton	—	—	338.00	—	753.00	—
5. Textile cloth	28494.08	—	25908.03	—	31569.78	—
6. Handloom machine	—	604.00	—	400.00	—	560.00
7. Band goods	775.00	—	490.00	—	447.00	—
8. Petroleum	—	91580.00	—	97698.00	—	95250.00
9. Coal	—	93798.00	—	87650.00	—	89768.00
10. Firewood & Fuel	—	219319.80	—	234395.29	—	194640.59
11. Timber	—	15854.00	—	16304.00	—	21151.10

Contd.

Appendix-1 — *Contd.*

Commodities	*1981*		*1982*		*1983*	
	Surplus	*Deficit*	*Surplus*	*Deficit*	*Surplus*	*Deficit*
12. Construction goods	—	470744.87	—	579803.86	—	733466.57
13. Fertilisers	—	188529.00	—	146873.00	—	215863.00
14. Sweets	67018.60	—	63293.80	—	66706.20	—
15. Food items	81587.43	—	334545.63	—	511307.01	—
16. Vegetables	11235.60	—	11976.00	—	12632.40	—
17. Fruits	3267.90	—	3541.70	—	3162.30	—
18. Livestock feed	8281.70	—	8119.70	—	5830.40	—
19. Crushed bone	1580.00	—	1574.00	—	1043.00	—
20. Leather	—	861.14	—	733.10	—	657.19
21. Matting	—	—	—	—	—	—
22. Milk	24137.01	—	18455.14	—	17601.45	—
23 Milk-made items	11015.36	—	14550.34	—	14875.85	—
24 Cutlery	72.64	—	68.92	—	85.38	—
25. Metal	—	24755.52	—	74701.78	—	72900.12
26. Railway material	300.00	—	—	38.00	—	507.00
27. Defence goods	—	28312.00	—	62788.00	—	97878.00
28. Other commodities	—	51599.49	—	53742.44	—	70277.00
Total	238723.46	1186665.82	485332.68	1355127.47	668002.61	1593089.57

APPENDIX 2-A

Meerut City – Places, Distance and the Amount of Passengers' Movement Through Railways – 1984

	Places	*Distance (in kilometres)*	*Passengers (in numbers)*
1.	Naurnagar	6	2540
2.	Partapur	7	16468
3.	Pavalikhas	9	4304
4.	Chandsara	10	6146
5.	Mohiuddinpur	12	42104
6.	Daurala	17	21116
7.	Kharkhauda	17	3178
8.	Modinagar	20	169843
9.	Kaili	22	2970
10.	Sakauti	25	19534
11.	Muradnagar	30	92649
12.	Hafizpur	30	528
13.	Hapur	30	40494
14.	Duhai	34	7718
15.	Khatauli	34	57392
16.	Guldhar	38	27177
17.	Jarauda Nara	43	1205
18.	Mansoorpur	43	10047
19.	Ghaziabad	48	209249
20.	Gulaothi	49	7624
21.	Sahibabad	55	24997
22.	Muzaffarnagar	56	87682
23.	Baral	56	1735
24.	Bamanheri	60	800
25.	Shahadara	62	188987
26.	Ruhana Kalan	68	2542
27.	Delhi	68	529864
28.	Bulandshahr	71	14034
29.	Maman	78	2108
30.	Deoband	80	18144
31.	Gajraula	81	387
32.	Talheri Bujurg	84	1168

Contd.

Appendix 2-A — *Contd.*

	Places	*Distance (in kilometres)*	*Passengers (in numbers)*
33.	Khurja	93	19141
34.	Nagal	97	14514
35.	Bahadurgarh	98	401
36.	Tapri	101	378
37.	Amroha	104	816
38.	Saharanpur	114	137989
39.	Rajpura	123	5871
40.	Moradabad	134	1738
41.	Aligarh	137	6948
42.	Jagadhari	144	12953
43.	Roorkee	149	3093
44.	Rampur	163	571
45.	Laksar	166	2174
46.	Hathras	167	838
47.	Hardwar	195	4586
48.	Ambala Cantt.	195	50727
49.	Mathura	213	1032
50.	Tundala	215	3110
51.	Bareilly	226	1235
52.	Firozabad	231	2890
53.	Dehradun	247	1944
54.	Sarindh	249	3786
55.	Sikohabad	251	6732
56.	Khanna	266	4006
57.	Agra	269	3876
58.	Murena	273	2231
59.	Shahjanpur	297	1737
60.	Ropar	298	5708
61.	Etawah	307	16192
62.	Ludhiana	309	17363
63.	Phillor	322	5212
64.	Kiratpur Sahib	325	3806
65.	Bharthana	327	628
66.	Phagwara	345	4579
67.	Farrukhabad	358	1178
68.	Phaphund	363	3876
69.	Gangapur City	366	1203
70.	Jalandhar Cantt.	366	24851
71.	Gwalior	385	350
72.	Vyas	402	1457
73.	Rura	403	782

Contd.

Appendix 2-A — *Contd.*

	Places	*Distance (in kilometres)*	*Passengers (in numbers)*
74.	Amritsar	444	25814
75.	Kanpur	445	109695
76.	Ajmer	446	724
77.	Lucknow	461	76623
78.	Atari Border	473	2138
79.	Pathankot	481	6444
80.	Fatehpur	521	2518
81.	Kota	537	4276
82.	Jammu Tawi	583	17568
83.	Gonda	586	4276
84.	Bharwari	601	2438
85.	Bina	636	288
86.	Allahabad	638	77220
87.	Basti	737	863
88.	Gorakhpur	740	2138
89.	Varanasi	747	4307
90.	Bhopal	773	326
91.	Mugal Sarai	791	1567
92.	Ratlam	804	338
93.	Bhatani	810	735
94.	Siwan	859	1270
95.	Indore	880	565
96.	Baroda	1065	968
97.	Nagpur	1162	1564
98.	Saharsa	1258	340
99.	Bombay	1457	10959

APPENDIX 2-B

Meerut City – Routes Distance and the Amount of Passengers' Movement Through Highways – 1984

	Routes	*Distance (in km)*	*Passengers (in numbers)*
1	Meerut - Rajpura	8.5	98671
2.	Meerut - Khatauli	13	385846
3.	Meerut - Bhola	16	402780
4.	Meerut - Maukhas	18	177125
5.	Meerut - Sardhana	22	434869
6.	Meerut - Lawar	24	107426
7.	Meerut - Sakauti	25	126504
8.	Meerut - Parikshhatgarh	28	98985
9.	Meerut - Niwari	29	165132
10.	Meerut - Faridnagar	33	131063
11.	Meerut - Hastinapur	39	454452
12.	Meerut - Garhmukteswar	44	434656
13.	Meerut - Ghaziabad	49	5966802
14.	Meerut - Budhana	63	460615
15.	Meerut - Delhi	76	2819117
16.	Meerut - Moradabad	124	5413
17.	Meerut - Aligarh	138	174500
18.	Meerut - Hardwar	141	1658595
19.	Meerut - Rampur	153	17098
20.	Meerut - Karnal	156	157916
21.	Meerut - Rishikesh	169	226461
22.	Meerut - Hathras	172	133406
23.	Meerut - Sikandrarao	177	16293
24.	Meerut - Dehradun	183	201063
25.	Meerut - Kashipur	187	30330
26.	Meerut - Mathura	199	213184
27.	Meerut - Rudrapur	201	35286
28.	Meerut - Kotdwar	204	141692
29.	Meerut - Etah	209	66098
30.	Meerut - Bareilly	215	242921
31.	Meerut - Agra	224	977318
32.	Meerut - Nagina	243	66278
33.	Meerut - Haldwani	244	182846
34.	Meerut - Chandigarh	278	89181
35.	Meerut - Tanakpur	306	144822
36.	Meerut - Farrukhabad	348	59214
37.	Meerut - Paliya	394	17239
38.	Meerut - Lakhimpur	402	29002
39.	Meerut - Lucknow	476	58272

Note. Routes given above included traffic movement to intervening places which vary from one to as many as ten.

APPENDIX 3-A

Technological Determinants in District Meerut – 1979

Blocks	Variable Numbers*					
	1	*2*	*3*	*4*	*5*	*6*
1. Meerut	213.85	905.40	448.39	4551.18	1910.01	497
2. Jani	230.98	535.27	592.67	1450.21	3879.16	38
3. Rohta	269.82	878.54	595.77	907.32	3208.06	16
4. Rajpura	316.85	1015.87	625.86	1554.84	3740.83	14
5. Kharkhauda	233.06	1060.30	312.13	1976.04	3029.87	177
6. Mawana	252.86	297.17	1280.88	2198.80	2394.80	24
7. Hastinapur	187.94	338.30	406.14	614.26	2569.05	27
8. Parikshhatgarh	226.07	489.32	879.68	1192.66	2353.99	—
9. Machhra	278.36	632.88	1183.21	1249.12	2226.62	32
10. Sardhana	240.93	741.95	542.41	1267.68	2242.76	—
11. Saroorpur	243.87	612.46	502.56	1263.20	4171.98	—
12. Binoli	159.27	1145.54	120.39	1274.61	2057.28	10
13. Daurala	313.43	713.55	581.38	1559.87	2099.97	10
14. Baghpat	252.52	880.64	768.24	2273.44	4723.83	14
15. Baraut	237.67	803.76	465.31	1644.05	5684.70	—
16. Chhaprauli	354.60	997.54	787.78	750.20	4790.15	—
17. Khekra	290.53	1093.69	343.87	635.99	3301.62	22
18. Pilana	191.16	943.35	381.24	464.63	6392.68	—

* The variable names may be seen in Chapter 6, section indicators.

APPENDIX 3-B

Transport Determinants in District Meerut – 1979

Blocks	Variable Numbers*							
	1	*2*	*3*	*4*	*5*	*6*	*7*	*8*
1. Meerut	527	80.52	22.57	2.51	11.28	1.25	8437.22	6.27
2. Jani	422	81.50	14.61	–	14.61	1.12	–	4.50
3. Rohta	189	40.75	17.66	–	26.49	1.26	–	–
4. Rajpura	260	44.33	13.10	–	13.10	0.93	–	1.87
5. Kharkhauda	176	45.75	20.07	2.68	10.70	1.34	–	1.34
6. Mawana	118	33.88	17.61	–	17.61	1.04	111.91	34.19
7. Hastinapur	202	94.35	14.45	–	28.89	1.44	15.89	1.44
8. Parikshhat-garh	98	27.20	13.10	–	22.16	1.01	24.18	10.07
9. Machhra	193	47.87	15.35	–	9.44	2.36	30.69	8.26
10. Sardhana	242	64.80	31.03	1.24	27.31	1.24	83.18	–
11. Saroorpur	155	38.09	13.25	–	19.87	2.21	–	13.25
12. Binoli	134	24.69	22.81	–	10.59	1.63	8.96	5.70
13. Daurala	150	35.99	11.51	1.05	10.46	4.18	24.07	26.16
14. Baghpat	234	44.00	10.97	–	8.78	2.19	34.02	28.53
15. Baraut	206	22.65	29.40	0.73	11.02	0.73	234.48	1.47
16. Chhaprauli	204	38.88	15.55	–	8.29	1.04	13.48	14.51
17. Khekra	269	49.62	11.25	–	12.38	1.12	11.25	16.88
18. Pilana	199	43.50	13.99	–	8.99	–	25.99	12.00

* The variable names may be seen in Chapter 6, section indicators.

Institutional Determinants in District Meerut – 1979

Blocks	Variable Numbers* 1	2	3	4	5	6	7	8	9	10	11	12	13	14	15	16
1 Meerut	6.15	185.19	1.23	304	3.08	0.96	—	11.73	0.58	250	0.58	2.65	28.27	2.50	27.68	1.35
2. Jani	2.36	254.00	0.84	148	2.90	0.91	1.45	3.64	3.45	175	1.27	0.25	57.87	2.36	30.35	1.82
3. Rohta	2.55	270.16	1.02	96	3.64	1.18	1.82	7.06	2.55	65	0.78	2.22	16.59	2.55	33.91	1.37
4. Rajpura	1.35	256.02	1.13	85	4.41	0.85	—	4.91	3.73	78	1.02	0.32	19.86	2.20	27.30	1.52
5. Kharkhauda	1.63	301.16	1.25	203	3.32	0.46	—	5.81	4.65	64	0.70	4.68	62.99	3.02	25.05	0.70
6. Mawana	5.17	324.46	1.02	202	—	1.38	1.55	12.24	4.14	368	1.90	4.40	76.73	2.24	31.70	2.07
7. Hastinapur	1.07	285.17	0.74	174	—	0.95	1.19	5.12	2.62	94	1.19	0.37	44.03	1.43	25.75	1.19
8. Parikshhatgarh	1.37	277.73	0.92	106	—	0.82	1.42	1.23	5.07	189	1.37	1.06	24.72	1.78	25.85	1.37
9. Machhra	1.96	344.41	0.83	149	4.88	2.61	2.44	11.30	6.52	125	1.96	0.55	38.49	2.83	27.65	2.17
10. Sardhana	5.00	364.24	0.76	178	2.33	1.09	4.65	42.83	5.22	96	1.52	0.79	48.86	2.83	31.20	2.17
11. Saroorpur	4.41	495.20	1.88	144	1.17	1.47	1.17	12.65	6.76	398	1.76	1.07	40.56	3.53	25.44	2.65
12. Binoli	4.07	349.20	2.20	111	1.65	1.69	1.65	16.10	1.35	99	0.51	1.00	17.87	2.54	28.40	0.85
13. Daurala	6.67	332.33	0.82	139	—	0.70	2.45	49.12	1.75	78	1.05	1.32	19.33	2.28	28.65	1.05
14. Baghpat	6.52	327.11	1.33	146	—	1.96	1.21	3.91	—	298	3.04	1.30	24.82	2.61	31.74	1.96
15. Baraut	4.53	328.87	3.85	85	—	1.51	0.57	61.32	3.21	197	1.13	0.96	20.15	3.02	30.28	1.51
16. Chhaprauli	10.74	636.33	1.96	112	2.26	3.70	1.13	156.66	7.41	188	2.22	0.71	39.88	4.07	30.24	2.59
17. Khekra	2.95	395.68	0.48	87	1.85	1.14	1.85	20.33	—	174	1.82	0.55	5.52	2.50	32.67	1.82
18. Pilana	5.09	318.94	0.64	116	0.71	1.32	0.71	1.51	2.26	245	1.51	0.51	19.81	2.26	27.68	1.51

* The variable names may be seen in Chapter 6, section indicators.

APPENDIX 3-D

Technological Determinants in District Meerut – 1984

Blocks	*Variable Numbers**					
	1	*2*	*3*	*4*	*5*	*6*
1. Meerut	230.16	1106.83	851.90	54.48	1921.53	47146.00
2. Jani	228.80	613.52	772.14	18.63	2341.70	68961.00
3. Rohta	272.14	1033.44	822.17	13.78	2489.93	22245.00
4. Rajpura	323.02	1154.22	766.86	19.98	2181.41	21830.00
5. Kharkhauda	238.92	1314.37	463.85	24.02	1927.72	23098.00
6. Mawana	338.74	455.81	1964.10	33.14	2721.20	17428.00
7. Hastinapur	186.60	425.36	597.12	8.99	1616.69	24168.00
8. Parikshhat-garh	220.20	553.91	1022.94	14.73	2310.37	19231.00
9. Machhra	279.22	766.29	1452.25	16.67	3249.25	18736.00
10. Sardhana	266.50	878.93	787.40	18.27	2401.28	43566.00
11. Saroorpur	262.15	835.08	716.16	17.80	2357.38	19606.00
12. Binoli	172.03	1333.36	203.43	16.56	2082.87	14911.00
13. Daurala	345.22	901.55	901.55	21.48	2861.54	282757.00
14. Baghpat	224.02	912.96	931.10	24.15	3060.08	36223.00
15. Baraut	306.19	1205.24	792.70	24.44	2487.09	26424.00
16. Chhaprauli	352.28	1167.99	1085.14	8.49	3794.24	104849.00
17. Khekra	290.55	1307.10	566.86	9.20	3287.34	23272.00
18. Pilana	199.60	1130.90	604.47	7.07	3066.47	16599.00

* The variable names may be seen in Chapter 6, section indicators.

APPENDIX 3-E

Transport Determinants in District Meerut – 1984

Blocks	*Variable Numbers*							
	1	*2*	*3*	*4*	*5*	*6*	*7*	*8*
1. Meerut	1386	142	13.21	4.40	9.91	2.20	53.96	3.30
2. Jani	464	80	13.31	–	12.36	0.95	3.80	1.90
3. Rohta	213	38	14.49	–	21.74	1.04	–	4.14
4. Rajpura	296	45	11.02	–	16.95	0.85	4.24	1.70
5. Khar-khauda	199	44	16.92	1.13	9.02	–	–	1.13
6. Mawana	474	128	15.40	–	16.36	–	4.81	2.89
7. Hastinapur	217	93	11.92	–	26.48	–	10.59	1.32
8. Parikshhat-garh	114	29	13.46	–	19.74	0.90	3.59	2.69
9. Machhra	241	52	13.89	–	8.55	1.07	6.41	2.14
10. Sardhana	274	62	14.38	1.03	22.60	2.05	4.11	2.05
11. Saroorpur	208	44	12.89	–	17.85	1.98	13.89	2.98
12. Binoli	202	28	17.50	–	8.75	–	4.04	1.35
13. Daurala	538	110	10.12	0.92	9.20	3.68	15.64	3.68
14. Baghpat	582	103	21.35	2.03	8.13	–	–	2.03
15. Baraut	229	41	25.09	3.14	9.41	0.63	2.51	1.25
16. Chhaprauli	236	44	17.91	–	7.96	–	–	1.00
17. Khekra	341	63	13.74	3.43	12.59	–	–	2.29
18. Pilana	240	45	13.50	–	7.60	–	1.69	1.69

* The variable names may be seen in Chapter 6, section indicators.

APPENDIX 3-F

Institutional Determinants in District Meerut – 1984

Blocks	Variable Numbers*															
	1	2	3	4	5	6	7	8	9	10	11	12	13	14	15	16
1. Meerut	2.68	234.88	3.34	308	3.59	0.98	17.97	6.10	2.93	315	0.73	3.08	49.29	4.15	28.92	1.70
2. Jani	2.32	249.46	4.05	111	4.80	0.89	16.82	2.50	4.29	201	1.25	6.04	36.00	3.21	31.49	1.78
3. Rohta	2.45	281.65	2.84	83	4.74	1.22	21.34	6.53	3.88	80	0.82	7.80	53.66	3.27	34.93	1.22
4. Rajpura	3.08	290.48	2.90	119	10.35	0.96	20.71	3.27	5.77	99	1.15	8.54	53.09	3.65	28.37	1.54
5. Kharkhauda	1.90	308.33	4.81	145	5.25	0.48	7.87	3.81	5.43	75	0.71	9.82	118.67	4.29	26.08	1.43
6. Mawana	3.16	330.16	2.63	208	—	1.40	13.33	5.61	6.49	419	1.93	15.31	76.72	3.16	32.71	2.11
7. Hastinapur	1.98	295.73	1.93	63	—	0.99	13.29	2.10	4.44	105	1.23	12.64	83.10	2.47	26.78	1.23
8. Parikshhatgarh	2.50	281.58	2.24	97	—	0.83	7.05	5.56	6.25	215	1.39	8.21	74.77	2.36	26.95	1.39
9. Machhra	3.33	330.06	3.69	100	1.75	2.50	27.94	6.25	7.71	135	1.88	2.66	65.17	3.75	28.68	2.08
10. Sardhana	3.26	364.24	3.17	121	2.40	1.09	16.77	5.22	4.13	105	1.52	8.52	85.73	3.70	32.23	2.17
11. Saroorpur	4.85	510.21	6.58	111	2.23	1.52	17.82	4.85	5.76	419	1.82	6.11	53.04	4.55	26.45	2.42
12. Binoli	1.72	355.22	3.47	104	1.79	1.55	21.45	5.86	8.28	120	0.52	2.00	45.26	3.28	30.42	0.69
13. Daurala	2.32	338.27	2.36	138	—	0.71	7.98	4.82	6.25	89	1.07	12.83	48.03	3.21	30.71	1.07
14. Baghpat	1.63	307.08	2.69	145	—	1.63	13.54	5.71	1.63	327	0.82	1.74	36.78	3.47	32.82	1.63
15. Baraut	3.33	322.78	6.48	150	1.72	1.67	12.05	9.44	6.48	205	1.11	1.41	26.04	3.70	30.43	1.67
16. Chhaprauli	5.19	636.33	5.74	52	4.29	3.70	20.01	7.41	10.00	200	2.22	7.16	84.92	5.56	30.27	2.59
17. Khekra	3.41	394.68	2.68	53	3.73	1.14	9.33	1.59	0.45	181	1.82	2.95	6.91	3.86	32.67	1.82
18. Pilana	3.13	352.17	1.73	41	1.58	1.46	14.23	1.67	4.58	260	1.67	2.15	37.88	3.54	28.70	1.88

* The variable names may be seen in Chapter 6, section indicators.

APPENDIX 4-A

Geographical Problems in District Meerut – 1984

Blocks	Variable Numbers													
	1	*2*	*3*	*4*	*5*	*6*	*7*	*8*	*9*	*10*	*11*	*12*	*13*	*14*
1. Meerut	5	5	9.42	—	31.71	4.88	2.44	90.24	56.10	2.44	17.07	17.07	24.39	87.80
2. Jani	16	16	2.51	3.57	98.21	8.93	16.07	60.71	62.50	—	17.86	17.86	12.50	76.78
3. Rohta	19	19	4.60	14.29	91.84	6.12	2.04	87.76	65.31	—	2.04	2.04	24.49	75.51
4. Rajpura	8	8	1.76	—	100.00	—	17.31	90.38	44.23	26.92	9.62	32.69	19.23	98.08
5. Kharkhauda	19	2	3.21	19.05	59.52	40.48	7.14	73.81	28.57	2.44	11.90	11.90	45.24	69.05
6. Mawana	26	26	2.26	3.51	100.00	19.30	10.53	85.96	75.44	33.33	10.53	12.28	24.56	82.46
7. Hastinapur	35	35	7.66	9.88	100.00	19.75	56.79	86.42	45.68	13.58	29.63	29.63	30.86	91.36
8. Parikshhatgarh	26	26	12.54	11.11	100.00	22.22	40.28	100.00	84.72	40.28	41.67	43.06	4.17	90.28
9. Machhra	23	23	1.12	4.17	100.00	8.33	2.08	54.17	68.75	12.50	—	10.42	—	68.75
10. Sardhana	20	12	2.76	—	63.04	8.70	10.87	60.87	47.83	—	8.70	8.70	4.35	78.26
11. Saroorpur	26	17	6.89	—	100.00	—	12.12	69.70	63.64	6.06	18.18	18.18	27.27	87.88
12. Binoli	41	11	4.52	15.52	96.55	25.86	10.34	77.59	53.45	20.69	20.69	24.14	50.00	81.03
13. Daurala	14	—	2.28	3.57	48.21	5.36	8.93	50.00	64.29	41.07	21.43	21.43	23.21	67.86
14. Baghpat	50	5	2.80	6.12	63.27	36.73	—	67.35	69.39	42.86	10.20	10.20	30.61	87.75
15. Baraut	53	—	3.49	20.37	31.48	11.11	—	46.30	44.44	—	11.11	12.96	11.11	66.67
16. Chhaprauli	65	11	4.56	—	88.89	7.41	11.11	62.96	48.15	—	—	—	7.41	74.07
17. Khekra	61	—	5.67	—	63.64	9.10	9.10	79.55	72.73	—	27.27	27.27	9.09	68.18
18. Pilana	36	13	2.72	—	93.75	35.42	—	35.42	56.25	45.83	6.25	6.25	6.25	81.25

* The variable names may be seen in Chapter 7, Part I, section indicators.

APPENDIX 4-B

Social Problems in District Meerut – 1984

Blocks	*Variable Numbers**							
	1	*2*	*3*	*4*	*5*	*6*	*7*	*8*
1. Meerut	13.92	654	28.92	15.82	5.71	2.49	0.15	45.50
2. Jani	18.21	518	31.49	14.65	3.20	4.98	0.29	—
3. Rohta	21.89	463	34.93	16.82	7.31	3.94	0.26	—
4. Rajpura	10.43	587	28.37	19.93	3.88	6.22	0.20	—
5. Kharkhauda	15.61	395	26.08	16.91	3.65	5.60	0.82	0.95
6. Mawana	7.65	348	32.71	21.03	7.31	7.68	0.32	5.69
7. Hastinapur	9.29	214	26.78	25.34	3.88	7.47	2.13	6.16
8. Parikshhat-garh	12.16	359	26.95	19.63	9.13	9.34	1.09	3.79
9. Machhra	10.49	404	28.68	13.43	6.85	7.68	0.77	1.90
10. Sardhana	20.85	373	32.23	37.62	5.48	3.94	0.59	4.74
11. Saroorpur	11.32	406	26.45	30.17	3.65	3.94	0.47	—
12. Binoli	5.78	621	30.42	25.84	7.76	9.96	0.52	0.47
13. Daurala	5.58	449	30.71	28.65	6.16	7.26	0.49	1.90
14. Baghpat	4.27	551	32.82	17.58	6.39	1.66	1.45	12.32
15. Baraut	17.20	487	30.43	15.97	11.64	7.26	2.44	12.79
16. Chhaprauli	4.18	524	30.27	19.52	4.57	5.60	2.84	1.42
17. Khekra	22.96	546	32.67	23.89	1.60	0.41	2.10	1.90
18. Pilana	23.19	441	28.70	22.56	1.83	4.56	1.07	0.47

Four indicators, viz., 3, 5, 6, 7 and 8, have been taken in their inverse form. For instance, illiterate population is taken from the inverse of literate population.

* The variable names may be seen in Chapter 7, Part I, section indicators.

APPENDIX 4-C

Economic Problems in District Meerut – 1984

Blocks	Variable numbers*											
	1	2	3	4	5	6	7	8	9	10	11	12
1. Meerut	4.44	6.88	87.80	100.00	60.98	17.07	43.90	87.80	41.46	90.24	99.42	0.58
2. Jani	5.55	2.01	100.00	69.64	92.86	—	25.00	100.00	14.28	100.00	97.27	2.73
3. Rohta	5.26	5.00	91.84	100.00	100.00	24.49	12.24	95.92	—	100.00	99.49	0.51
4. Rajpura	5.64	8.12	75.00	57.69	69.23	15.38	53.85	100.00	7.69	82.69	99.59	0.41
5. Kharkhauda	6.20	2.25	69.05	88.10	80.95	30.95	42.86	95.24	21.43	100.00	99.91	0.09
6. Mawana	6.10	10.22	84.21	100.00	84.21	14.04	73.68	77.19	22.81	82.46	99.06	0.94
7. Hastinapur	4.86	2.69	97.53	83.95	97.53	19.75	51.85	97.53	29.63	97.53	85.03	14.97
8. Parikshhatgarh	5.86	4.56	93.06	91.67	100.00	37.50	61.11	88.89	44.44	90.28	98.43	1.57
9. Machhra	5.75	2.75	89.58	100.00	100.00	—	8.33	89.58	—	95.83	98.33	1.67
10. Sardhana	5.04	15.18	76.09	91.30	78.26	15.22	32.61	78.26	26.08	78.26	98.94	1.06
11. Saroorpur	4.77	1.22	96.97	93.94	96.97	21.21	81.82	96.97	27.27	96.97	98.30	1.70
12. Binoli	7.10	3.10	82.76	79.31	100.00	53.45	62.07	93.10	50.00	94.83	96.17	3.83
13. Daurala	5.17	4.20	96.43	92.86	66.07	21.43	28.57	100.00	1.78	100.00	99.89	0.11
14. Baghpat	5.62	8.12	100.00	87.76	100.00	34.69	38.77	67.35	28.57	87.75	92.01	7.99
15. Baraut	7.06	16.95	66.67	66.67	66.67	33.33	66.67	66.67	35.18	100.00	98.71	1.29
16. Chhaprauli	5.35	1.26	96.30	100.00	100.00	7.41	66.67	62.96	25.92	100.00	98.65	1.35
17. Khekra	4.07	4.23	68.18	100.00	100.00	13.64	63.64	68.18	13.64	100.00	93.51	6.49
18. Pilana	6.16	1.26	100.00	100.00	100.00	16.67	81.25	100.00	6.25	100.00	96.21	3.79

Two indicators, viz., 2 and 11, have been taken in their inverse form. For instance, proportion of industrial workers to represent the dispersal index of industrial workers.

* The variable names may be seen in Chapter 7, Part I, section indicators.

APPENDIX 5-A

Non-Agricultural Development in District Meerut – 1984

Blocks	*Variable numbers**					
	1	*2*	*3*	*4*	*5*	*6*
1. Meerut	1.55	13	2.20	1.22	25.90	25.90
2. Jani	1.31	13	9.64	1.07	26.75	21.12
3. Rohta	1.00	21	9.59	1.84	16.76	32.19
4. Rajpura	0.81	20	7.50	1.92	16.53	33.06
5. Kharkhauda	0.86	9	9.05	4.05	25.00	25.00
6. Mawana	1.00	17	5.09	2.28	25.00	25.00
7. Hastinapur	0.92	20	0.99	2.59	24.95	25.14
8. Parikshhatgarh	0.80	22	4.17	1.11	25.00	25.00
9. Machhra	1.00	8	5.83	3.75	19.83	30.58
10. Sardhana	0.91	22	15.65	1.96	25.00	29.08
11. Saroorpur	0.83	18	20.00	4.85	21.00	29.50
12. Binoli	0.93	13	8.62	1.90	34.51	28.79
13. Daurala	0.69	11	4.64	0.71	24.17	24.17
14. Baghpat	0.87	10	5.31	0.41	23.52	24.83
15. Baraut	0.87	20	7.59	0.19	25.00	25.00
16. Chhaprauli	1.00	8	30.37	2.96	19.91	35.35
17. Khekra	0.93	14	3.18	0.23	27.54	28.26
18. Pilana	0.97	9	7.08	0.42	23.74	25.80

* The variable names may be seen in Chapter 7, Part II, section indicators.

APPENDIX 5-B

Agricultural Development in District Meerut – 1984

Blocks	Variable numbers*					
	1	2	3	4	5	6
1. Meerut	25.10	47.37	0.73	2125	19.10	13.09
2. Jani	30.25	46.56	—	2466	8.29	4.07
3. Rohta	35.06	45.18	0.91	3386	25.69	4.56
4. Rajpura	33.72	45.58	0.18	2566	11.29	4.76
5. Kharkhauda	31.70	50.37	4.73	3074	12.91	9.25
6. Mawana	44.96	34.02	—	3341	20.35	4.22
7. Hastinapur	37.54	43.67	0.56	4872	15.73	5.79
8. Parikshhatgarh	38.34	38.40	—	4361	24.58	7.00
9. Machhra	32.74	43.66	—	4260	22.31	18.63
10. Sardhana	38.17	44.70	3.64	3291	18.92	6.64
11. Saroorpur	33.68	51.74	6.37	3188	11.36	6.60
12. Binoli	32.30	49.06	71.96	2630	15.98	18.51
13. Daurala	35.59	43.63	8.56	3117	16.78	2.59
14. Baghpat	28.15	49.94	0.18	3793	17.98	15.35
15. Baraut	29.01	42.75	1.27	2661	23.37	13.56
16. Chhaprauli	44.62	40.31	0.36	4150	11.86	4.42
17. Khekra	14.44	60.25	0.55	3761	4.91	0.47
18. Pilana	21.11	55.79	—	3866	4.89	11.88

* The variable names may be seen in Chapter 7, Part II, section indicators.

APPENDIX 5-C

Integrated Rural Development Programme and Poverty Eradication in District Meerut – 1984

Blocks	*Variable numbers**								
	1	*2*	*3*	*4*	*5*	*6*	*7*	*8*	*9*
1. Meerut	24.58	7.33	1585.72	5.31	36.11	6.15	2.37	2.65	4.43
2. Jani	23.10	0.89	3.80	2.64	69.25	7.84	10.16	9.87	8.32
3. Rohta	22.93	3.21	8.28	4.21	50.75	7.33	7.67	8.72	7.06
4. Rajpura	24.33	0.40	6.78	4.02	50.78	4.99	10.72	1.58	3.26
5. Kharkhauda	24.91	0.16	4.51	6.13	30.22	4.05	0.32	0.34	3.44
6. Mawana	26.53	0.76	27.91	4.83	42.73	6.76	0.32	0.34	3.44
7. Hastinapur	25.69	2.40	31.77	4.86	42.52	10.94	7.67	8.72	7.06
8. Parikshhatgarh	30.81	0.61	3.58	4.67	42.21	4.91	0.32	0.34	3.44
9. Machhra	19.55	0.37	4.27	4.50	50.28	7.42	5.14	5.90	5.83
10. Sardhana	15.74	0.79	8.21	5.99	29.01	0.80	0.32	0.34	3.44
11. Saroorpur	24.89	0.21	3.97	4.93	40.63	4.08	2.37	2.65	4.43
12. Binoli	24.98	0.28	5.38	2.14	80.57	4.91	9.84	9.53	8.15
13. Daurala	32.59	0.25	7.36	2.83	67.98	4.14	9.84	11.24	8.15
14. Baghpat	25.31	1.67	8.13	2.92	65.43	1.30	9.84	11.24	8.15
15. Baraut	23.23	1.24	5.02	3.48	56.74	5.09	0.32	0.34	0.17
16. Chhaprauli	26.12	1.07	3.98	2.63	73.25	8.21	10.08	11.49	8.28
17. Khekra	26.70	0.35	4.58	4.84	41.69	2.41	2.33	2.86	4.52
18. Pilana	22.44	0.55	3.38	2.65	81.98	1.70	10.36	11.84	8.43

* The variable names may be seen in Chapter 7, Part II, section indicators.

APPENDIX 5-D

Budget Allocation in Annual Plan in District Meerut – 1987

Blocks	*Variable numbers**						
	1	*2*	*3*	*4*	*5*	*6*	*7*
1. Meerut	5880	39924	36533	120	768	325	83550
2. Jani	5512	2455	1745	125	40	80	9957
3. Rohta	5535	1635	1211	100	40	50	8571
4. Rajpura	5590	2930	1996	100	40	72	10728
5. Kharkhauda	5555	2060	1743	100	48	80	9586
6. Mawana	5552	3695	3102	100	64	130	12643
7. Hastinapur	5520	3215	2380	125	64	100	11404
8. Parikshhatgarh	5557	2530	1740	125	56	72	10080
9. Machhra	5552	2990	2070	125	64	60	10861
10. Sardhana	5565	3555	3684	125	80	100	13119
11. Saroorpur	5515	2730	1510	125	64	70	10014
12. Binoli	5555	3330	2140	125	80	70	11270
13. Daurala	5555	3805	1780	125	64	40	11369
14. Baghpat	5570	3870	3560	100	64	104	13268
15. Baraut	5360	4735	4741	105	112	86	15139
16. Chhaprauli	5557	2670	1803	125	48	45	10248
17. Khekra	5515	3270	1190	125	40	60	10200
18. Pilana	5555	3468	1640	125	40	60	10888

* The variable names may be seen in Chapter 7, Part II, section indicators.

BIBLIOGRAPHY

Adalemo, I.A., "Traders Travel Patterns : Marketing Rings and Pattern of Market Shifts", *Nigerian Geographical Journal*, 18 (1975).

Alam, S. Manzoor and Khan, Waheeduddin, *Metropolitan Hyderabad and its Region : A Strategy for Development*, Asia Publishing House, Bombay (1972).

Alexander, J.W. *et al.*, "Freight Rates : Selected Aspects of Uniform and Nodal Regions", *Economic Geography*, 34, (1958), pp. 1-18.

Archer, J.E. and Dalton, T.H., *Fieldwork in Geography*, Ch. 6, Batsford (1968).

Baer, W. and Melissa Birch, "The International Economic Relations of a Small Country : The Case of Paraguay", *Economic Development and Cultural Change*, 35 (3), The University of Chicago Press, (April 1987), pp. 601-27.

Banerjee, B., *Rural to Urban Migration and the Urban Labour Market – A Case Study of Delhi*, Himalaya Publishing House, Delhi, (1987), pp. 1-279.

Berry, B.J.L., "Essays on Commodity Flows and the Spatial Structure of the Indian Economy", *Research Paper No. 109*, University of Chicago, Department of Geography (1966).

Berry, B.J.L., Barnum, H.G. and Tennant, R.J., "Retail Location and Consumer Behaviour", *Papers and Proceedings of the Regional Science Association*, (1962), pp. 65-106.

Borgers, A. and Harry Timmermans, "A Model of Pedestrian Route Choice and Demand for Retail Facilities within Inner City Shopping Areas", *Geographical Analysis*, 18 (2), Ohio State University Press, Ohio, (April 1986), pp. 115-28.

Bose, P.K. and Bhattacharya, R.N., "Growth of Mysore City and its Passenger Transport", *Geographical Review of India*, 46 (4) (1984), pp. 34-45.

Bromley, R.J., *Periodic Markets, Daily Markets and Fairs : A Bibliography*, Melbourne : Manash Publications in Geography, (10) (1974).

Central Statistical Organisation, "Annual Survey of Industries 1959 to 1965", Vol. 1, Deptt. of Cabinet Secretariat, Government of India, Calcutta.

Census of India, Vol. 1, India, "Occupational Structure and Levels of Economic Development in India – A Regional Analysis" by Joseph E. Schwarzberg, *Monograph No. 4* (1961).

Chapman, M., *Population Movement in Tribal Society: The Case of Duidui and Pichahila, British Salomon Islands*, Ph.D. Dissertation, University of Washington (1970).

Chatterjee, A.B., "Nature of Commodity Flow of a Market in the Fringe Area of Calcutta", *Geographical Review of India*, 31 (3), Calcutta (1969).

Chisholm, M., "Freight Transport Costs, Industrial Location and Regional Development", in M. Chisholm and G. Manners (eds.), *Progress in Geography* (1971).

Chisholm, M. and Sullivan, P.O., *Freight Flows and Spatial Aspects of the British Economy*, Cambridge University Press (1973).

Clark, W.A.V., "Consumer Travel Patterns and the Concept of Range", *Annals of the Association of American Geographers*, 58, (1968), pp. 386-96.

Clark, W.A.V., "Spatial Structure of Retail Functions in a Newzealand City", *Newzealand Geographer*, 23, (1967), pp. 23-33.

Cooley, C.H., "Theory of Transportation" in *Transportation Geography : Comments and Readings*, M.E. Eliot Hurst (eds.) McGraw-Hill, (1974), p. 15.

Cooper, L., "The Transportation – Location Problem", *Operations Research*, 20 (1972).

Coppock, J.T., "Crops, Livestock and Enterprise Combinations in England and Wales", *Economic Geography*, 40, (1971), pp. 65-81.

Deshpande, C.D., Arunachalam, B. and Bhat, L.S., "South Kolaba - A Study in Settlement Hierarchy and Rural-Urban Interactions", *Mimeographed Report of an I.C.S.S.R. Sponsored Project* (New Delhi) (1979).

Eiichi, M., "The Traffic Revolution : Remarking the Export Sales System in China, 1866-1875", *Modern China*, 12 (1), London, (England), (Jan. 1986), pp. 75-102.

Eliot Hurst, M.E., *Transportation Geography : Comments and Readings*, McGraw-Hill, (1974), p. 9.

Garrison, W.L., "Connectivity of the Inter-State Highway System", *Papers and Proceedings of the Regional Science Association*, 6, (1960), pp. 121-37.

Genann, G.A., "Periodic Markets and Itinerant Traders Commodities in Lushoto, Tanzania" in Robert H.T. Smith, edited, *Internal Exchange Systems : Periodic Markets and Hawkers in Africa, Asia and Latin America*, Melbourne : Sorrett (1975).

Gillespie, A.E. and Goddard, J.B., "Advanced Telecommunication and Regional Economic Development", *The Geographical Journal*, 152 (3), (Nov. 1986), pp. 383-97.

Goddard, J.B., "Functional Regions within a City Centre: A Study by Factor Analysis of Taxi Flows in Central London", *Transactions of the Institute of British Geographers*, 49, (1970), pp. 161-80.

Godlund, S., "Bus Services in Sweden", *Lund Studies in Geography*, Series B (17) (1956).

Goode, W.J. and Hatt, P.K., *Methods in Social Research*, McGraw-Hill (1952).

Gopi, K.N., *Urban Growth and Industrial Locations*, Oxford and IBH Publishing Company, New Delhi (1980).

Gopi, K.N., *Process of Urban Fringe Development : A Model*, Concept Publishing Company, New Delhi (1978).

Gregory, S., *Statistical Methods and the Geographer*, Ch. 7, Arnold (1965).

Guldmann, J.M., "Interactions between Weather Stochasticity and the Locations of Pollution Sources and Receptors in Air Quality Planning: A Chance – Constrained Approach", *Geographical Analysis*, 18 (3), Ohio State University Press, Ohio (July 1986).

Haggett, P., *Locational Analysis in Human Geography*, Ch. 5, Arnold (1965).

Harvey, M.E., Hocking, R.T. and Brown, J.R., "The Chromatic Travelling – Salesman Problem and Its Application to Planning and Structuring Geographic Space", *Geographical Analysis*, 6, (1974), pp. 33-52.

Hay, A., *Transport for the Space Economy: A Geographical Study*, Macmillan (1973)

Hay, A.M. and R.H.T. Smith, *Inter-regional Trade and Money Flows in Nigeria*, Oxford University Press (1970).

Hodgson, M.J., "Towards More Realistic Allocation in Location-Allocation Models: An Interaction Approach", *Environment and Planning*, 24, (1978), pp. 65-85.

Hodder, B.W. and Ukwu U.I., *Markets in West Africa : Studies of Markets and Trade Among the Yoruba and Ibo*, Ibadan University Press (1969).

Holmes, J., Williams, F.B. and Brown, L.A., "Facility Location under a Maximum Travel Restriction : An Example Using Daycare Facilities", *Geographical Analysis*, 4, (1972), pp. 258-66.

Huff, James O. and Susan Hanson, "Repetition and Variability in Urban Travel", *Geographical Analysis*, 18 (2), Ohio State University Press, (April 1986), pp. 97-114.

Ilori, C.O., "Economic Study of Production and Distribution of Staple Food crops in Western Nigeria, Palo Alto" : Unpublished Ph.D. Dissertation, Standford University (1968).

Isard, W., *Location and Space Economy*, M.I.T. Press (1965).

Jackson, R.T., "Commuter Mining and the Kidston Gold Mine : Goodbye to Mining Towns?", *Geographical Journal of the Geographical Association*, 72 (2), 315, (Apr. 1987), pp. 162-64.

Kansky, K.J., "Structure of Transport Networks", *Research Paper No. 84, University of Chicago, Department of Geography* (1963).

Kansky, K.J., "Structure of Transportation Networks : Relationships between Network Geometry and Regional Characteristics", *Research Paper No. 84*, University of Chicago, Department of Geography (1963).

Kara, P.K., Tripathy S.N. and Singh, D.K., "A Comparative Analysis of Major Ports of India and their Traffic Characteristics", *NGJI*, 28 (1 and 2), Varanasi, (1982), pp. 60-66.

Kelly, M.E.O., "Activity Levels at Hub Facilities in Interacting Networks", *Geographical Analysis*, 18 (4), (Oct. 1986), pp. 343-56.

Kulkarni, K.M., *Urban Structure and Interaction : A Study of Nasik City Region*, Concept Publishing Co., New Delhi (1981).

Kulkarni, K.M., *Geography of Crowding and Human Response: A Study of Ahmedabad*, Concept Publishing Co., New Delhi (1984).

Kumar, N., "Economic Interdependence and Human Interaction in Meerut City", *Proceedings of the "Sixth N.A.G.I."*, Seminar, I.I.T. Kharagpur, 9-11 May 1985, pp. 137-38.

Kumar, N., "Determinants of Textile Production in India", *Proceedings of the Seminar on*, "Role of Textiles in National Development", Meerut College, Meerut, 16-17 March 1986, pp. 17-18.

Kumar, N., "Flow Links of Partapur Dairy in Meerut City", *Proceedings of the National Seminar on Industrial Development*, Policies and Planning in India, Tilak College, Auraiya-Etawah, 23-25 May 1986, pp. 1-3.

Kumar, N., "Nature of Rural Urban Interaction in Meerut", *Proceedings of the "8th N.A.G.I."* Seminar at University of Kashmir, Srinagar, 1-4 Oct 1986.

Kumar, N., "Commodity Flows and Human Interaction on Meerut-Delhi Axis", Paper Presented in the Seminar on *"N.C.R. Prospects and Problems"* at M.M.H. College, Ghaziabad, 8-9 November, (1986).

Kumar, N., "Production, Flows and Hierarchic Relation of Meerut City: A Case Study of Cotton Textiles", *The Geographical Observer*, 22, Meerut, (1986), pp. 32-45.

Kumar, N., "Agricultural Development and Interaction in Meerut", Paper Presented in the Seminar on "Rural Development Planning in India" at S.S.V. College, Hapur, 1-3 March, 1987.

Linneman, H., *An Econometric Study of International Trade Flows*, North Holland (1966).

Mitchell, J.C., "Wage Labour and African Population Movements in Central Africa" in *Essays on African Population*, Barbour, K.M. and Prothero, R.M. (eds.), London: Routledge (1961).

Mitchell, J.C., "Wage-Labour Mobility as Circulation" Paper Presented at the International Seminar on the Cross-Cultural Study of Circulation, East-West Population Institute, Hawaii (1978).

Monckhouse, F.J. and Wilkinson, H.R., *Maps and Diagrams : Their Compilation and Construction*, Methuen, 3rd edn. (1971).

Morrill, R.L., "The Shape of Diffusion in Space and Time", *Economic Geography*, 46 (Supplement), (1970), pp. 259-68.

Morrill, R.L., "Migration and Growth of Urban Settlements", *Lund Studies in Geography*, Series B, Human Geography, 22 (26), (1965), pp. 130-70.

Mortimore, M., "An 'Alternative Opportunity' Framework for Population Mobility, Illustrated from Nigeria" in *Population Distribution and Re-distribution in Africa*, J.I. Clarke and L.A. Kosinski (eds.) Heinemann, London (1981).

Murdie, R.A., "Cultural Differences in Consumer Travel", *Economic Geography*, 41, (1965), pp. 211-33.

N.C.A.E.R., *Traffic Survey of Madras Port*, New Delhi (1968).

N.C.A.E.R., *Traffic Survey of Cochin Port*, New Delhi (1969).

N.C.A.E.R., *Traffic Survey of Visakhapatnam Port*, New Delhi (1970).

Olsson, Gunnar, *Distance and Human Interaction : A Review and Bibliography*, Bibliography Series, No. 2, Regional Science Association Institute, Philadelphia Press (1965).

Oppenheim, A.N., "Questionnaire, Design and Attitude Measurement", Heinemann (1966).

Pal, M.N., "Regional Demand Projections for Different Commodities in India 1970-71 and 1975-76", *Indian Economic Journal*, 14 (4) (1977).

Pankaj, T., "A Study of the Hinterland Limits and Traffic Flow Patterns of the Port of Cochin", *Arthavijnana*, 10 (1968).

Prakasa Rao, V.L.S. and Tewari, V.K., *The Structure of an Indian Metropolis : Bangalore – An Urban Survey*, Allied Publishers, New Delhi (1980).

Pred, A.R. and Tornqvist, G.E., "Systems of Cities and Information Flows", *Lund Studies in Geography*, Ser. B, (38) (1973).

Ramu, G.N. and Wiebe, P., "Occupational and Educational Mobility in Relation to Caste in Urban India", *Sociology and Social Research*, 58 (1) (1973).

Rao, K.P.C., "Structural Changes, Sectoral Performance and Inter-Sectoral Linkages in Andhra Pradesh Economy", *MARGIN*, 18 (2), (Jan. 1986), pp. 62-69.

Raza, M., "Structure of Commodity Flows in India", Project Report Submitted to the *Planning Commission of India* (1980).

Raza, M. and Premi, K., "Indicators of Equity in Education : A Conceptual Framework", *Journal of Educational Planning and Administration*, I (2), (April 1987), pp. 1-29.

Raza, M., Aggarwal, Y.P. and Dutta, M., "A Spatial Analysis of Metropolitan Freight Flows in India", *Geog. Rev. Ind.*, 42 (4), (1980), pp. 323-41.

Reed, W.E., "Areal Interaction in India : Commodity Flows of the Bengal-Bihar Industrial Area", University of Chicago, Department of Geography, Research Paper No. 110 (1967).

Rezvan, E. and Engleburt, E.S., "Inter-Regional Linkages", University of California, Berkely, (1975), p. 3.

Robertson, K.A., "Pedestrian Skywalks : The Esthetic Impact on Down Town", *Landscape*, 29 (3), Berkley, California, (1987), pp. 43-47.

Sharp, C., *The Problem of Transport*, Pergamon Press (1965).

Smith, D.M., *Human Geography : A Welfare Approach*, Arnold Hienemann, London (1977).

Sopher, David E., Place and Landscape in India Tradition, *Landscape*, 29 (2), (1966), pp. 1-9.

Singh, I.D., "Pattern of Rail Traffic Flow in Rajasthan", *National Geographical Journal of India*, 17 (2&3), Varanasi, (1971), pp. 90-98.

Symanski, R., "God, Food, and Consumers in Periodic Market Systems", *Proceedings of the Association of American Geographers*, *5*, (1973), pp. 262-69.

Taaffe, E.J. and Howard, L.G., *Geography of Transportation*, (1973), pp. 107, 110 and 111.

Tornqvist, G., "Flows of Information and the Location of Economic Activities", *Geografiska Annaler*, Ser. B, 50, (1968), pp. 99-107.

Ullman, E.L., "Transportation Geography" in *American Geography : Inventory and Prospects*, P.E. James and C.L. Jones (eds.), Syracuse University Press (1956).

Ullman, E.L., "The Rail-Road Pattern of the United States", *Geographical Review*, 39, (1949), pp. 242-56.

Wanmali, S., *Regional Planning for Social Facilities*, National Institute of Rural Development, Hyderabad (1971).

Wanmali, S. and Ghosh, Avijit, "Pattern of Distribution of Consumer Goods in Rural India", *Management and Labour Studies*, 1, (1975), pp. 79-94.

Wanmali, S. and Ghosh, Avijit, "Distribution of Goods and Services in South Bihar : Patterns and Prospects", in Hauser, Waltar and Manor, James (eds.) *Two Faces of India, Social and Political Change in Bihar and Karnataka*, Oxford University Press, New Delhi (1975).

Wilson, A.G., "A Family of Spatial Interaction Models, and Associated Developments", *Environment and Planning*, 3, (1971), pp. 1-32.

Wolfe, R.I., "Effects of Ribbon Development on Traffic Flow", *Traffic Quarterly*, Jan., (1964), pp. 105-17.

Young, P.V., *Scientific Social Surveys and Research*, Prentice - Hall (1966)

INDEX